Deerfield Public Library
920 Waukegan Road
Deerfield, IL 60015

DEERFIELD PUBLIC LIBRARY

3 9094 32080 3230

WITHDRAWN

Analytical Skills for AI and Data Science

Building Skills for an AI-Driven Enterprise

D1264944

Daniel Vaughan

Beijing · Boston · Farnham · Sebastopol · Tokyo    O'REILLY®

Analytical Skills for AI and Data Science

by Daniel Vaughan

Copyright © 2020 Daniel Vaughan. All rights reserved.

Published by O'Reilly Media, Inc., 1005 Gravenstein Highway North, Sebastopol, CA 95472.

O'Reilly books may be purchased for educational, business, or sales promotional use. Online editions are also available for most titles (*http://oreilly.com*). For more information, contact our corporate/institutional sales department: 800-998-9938 or *corporate@oreilly.com*.

Acquisitions Editor: Jonathan Hassell
Development Editor: Michele Cronin
Production Editor: Daniel Elfanbaum
Copyeditor: Piper Editorial
Proofreader: Justin Billing

Indexer: Sue Klefstad
Interior Designer: David Futato
Cover Designer: Karen Montgomery
Illustrator: Rebecca Demarest

May 2020:   First Edition

Revision History for the First Edition
2020-05-21:   First Release

See *http://oreilly.com/catalog/errata.csp?isbn=9781492060949* for release details.

The O'Reilly logo is a registered trademark of O'Reilly Media, Inc. *Analytical Skills for AI and Data Science*, the cover image, and related trade dress are trademarks of O'Reilly Media, Inc.

The views expressed in this work are those of the author, and do not represent the publisher's views. While the publisher and the author have used good faith efforts to ensure that the information and instructions contained in this work are accurate, the publisher and the author disclaim all responsibility for errors or omissions, including without limitation responsibility for damages resulting from the use of or reliance on this work. Use of the information and instructions contained in this work is at your own risk. If any code samples or other technology this work contains or describes is subject to open source licenses or the intellectual property rights of others, it is your responsibility to ensure that your use thereof complies with such licenses and/or rights.

978-1-492-06094-9

[LSI]

Table of Contents

Preface

Why Analytical Skills for AI?

Judging from the headlines and commentary in social media during the second half of the 2010s, the age of artificial intelligence has finally arrived with its promises of automation and value creation. Not too long ago, a similar promise came with the big data revolution that started around 2005. And while it is true that some select companies have been able to disrupt industries through AI- and data-driven business models, many have yet to realize the promises.

There are several explanations for this lack of measurable results—all with some validity, surely—but the one put forward in this book is the general lack of analytical skills that are *complementary* to these new technologies.

The central premise of the book is that value at the enterprise is created by *making decisions*, not with data or predictive technologies alone. Nonetheless, we can piggyback on the big data and AI revolutions and start making better choices in a systematic and scalable way, by transforming our companies into modern AI- and data-driven decision-making enterprises.

To make better decisions, we first need to ask the right questions, forcing us to move from descriptive and predictive analyses to *prescriptive* courses of action. I will devote the first few chapters to clarifying these concepts and explaining how to ask better business questions suitable for this type of analysis. I will then delve into the anatomy of decision-making, starting with the consequences or outcomes we want to achieve, moving backward to the actions we can take, and discussing the problems and opportunities created by intervening uncertainty and causality. Finally, we will learn how to pose and solve prescriptive problems.

Use Case-Driven Approach

Since my aim is to help practitioners to create value from AI and data science using this analytical skillset, each chapter will illustrate how each skill works with the help of a collection of use cases. These were selected because I've found them valuable at work, because of their generality across industries, because students found them particularly interesting or useful, or because they are important building blocks for more complex problems commonly found in the industry. But in the end this choice was subjective, and depending on your industry, they may be more or less relevant.

What This Book Isn't

This book isn't about artificial intelligence or machine learning. This book is about the extra skills needed to be successful at creating value from these predictive technologies.

I have provided an introduction to machine learning in the Appendix for the purpose of being self-contained, but it isn't a detailed presentation of machine learning-related material nor was it planned as one. For that, you can consult many of the great books out there (some are mentioned in the Further Reading section of the Appendix).

Who This Book Is For

This book is for anyone wanting to create value from machine learning. I've used parts of the material with business students, data scientists, and business people alike.

The most advanced material deals with decision-making under uncertainty and optimization, so having a background in probability, statistics, or calculus will definitely help. For readers without this background, I've tried to make the presentation self-contained. On a first pass, you might just skip the technical details and focus on developing an intuition and an understanding of the main messages for each chapter.

- If you're a *business person* with no interest whatsoever in doing machine learning yourself, this book should at least help redirect the questions you want your data scientists to answer. Business people have great ideas, but they may have difficulty expressing what they want to more technical types. If you want to start using AI in your own line of work, this book will help you formulate and translate the questions so that others can work on the solutions. My hope is that it will also serve as inspiration to solve new problems you didn't think were resolvable.

- If you're a *data scientist*, this book will provide a holistic view of how you can approach your stakeholders and generate ideas to apply your technical knowledge. In my experience, data scientists become really good at solving predictive problems, but many times have difficulties delivering prescriptive courses of action. The result is that your work doesn't create as much value as you want and expect. If you've felt frustrated because your stakeholders don't understand the relevance of machine learning, this book could help you transform the question you're solving to take it "closer to the business."

- If you're neither one of these, the fact that you find the title compelling indicates that you have an interest in AI. Please recall the disclaimer in the previous section: *you won't learn to develop AI solutions in this book*. My aim is to help you translate business questions into prescriptive solutions using AI as an input.

What's Needed

I wrote this book in a style that is supposed to be readable for very different audiences. I do *not* expect the reader to have any prior knowledge of probability or statistics, machine learning, economics, or the theory of decision making.

Readers with such backgrounds will find the more technical material introductory, and that's actually great. In my opinion, the key to creating value through these techniques is to focus on the business question and not on the technical details. I hope that by focusing on the use cases you can find many new ways to solve the problems you're facing.

For readers with no background in these topics I've tried to provide a very minimal introduction to the key themes that I need to develop each of the use cases. If you're interested in going deeper, I've also provided a list of references that I've found useful, but I'm sure you can find many more on the internet. If you're not interested in going deeper, that's fine too. My advice is to focus on the broader picture and strengthen your intuition. That way you'll be able to ask the right questions to the right people at your companies.

What's really needed to get the most value from this book is curiosity. And if you've reached this paragraph, you're most likely well-equipped on that front.

Conventions Used in This Book

The following typographical conventions are used in this book:

Italic
> Indicates new terms, URLs, email addresses, filenames, and file extensions.

`Constant width`
> Used for program listings, as well as within paragraphs to refer to program elements such as variable or function names, databases, data types, environment variables, statements, and keywords.

`Constant width bold`
Shows commands or other text that should be typed literally by the user.

`Constant width italic`
Shows text that should be replaced with user-supplied values or by values determined by context.

This element signifies a tip or suggestion.

This element signifies a general note.

This element indicates a warning or caution.

Using Code Examples

Supplemental material (code examples, exercises, etc.) is available for download at *https://github.com/dvaughan79/analyticalskillsbook*.

This book is here to help you get your job done. In general, if example code is offered with this book, you may use it in your programs and documentation. You do not need to contact us for permission unless you're reproducing a significant portion of the code. For example, writing a program that uses several chunks of code from this book does not require permission. Selling or distributing examples from O'Reilly books does require permission. Answering a question by citing this book and quoting example code does not require permission. Incorporating a significant amount of example code from this book into your product's documentation does require permission.

We appreciate, but do not require, attribution. An attribution usually includes the title, author, publisher, and ISBN. For example: "*Analytical Skills for AI and Data Science* by Daniel Vaughan (O'Reilly). Copyright 2020 Daniel Vaughan, 978-1-492-06094-9."

If you feel your use of code examples falls outside fair use or the permission given above, feel free to contact us at *permissions@oreilly.com*.

O'Reilly Online Learning

 For almost 40 years, *O'Reilly Media* has provided technology and business training, knowledge, and insight to help companies succeed.

Our unique network of experts and innovators share their knowledge and expertise through books, articles, and our online learning platform. O'Reilly's online learning platform gives you on-demand access to live training courses, in-depth learning paths, interactive coding environments, and a vast collection of text and video from O'Reilly and 200+ other publishers. For more information, please visit *http://oreilly.com*.

How to Contact Us

Please address comments and questions concerning this book to the publisher:

> O'Reilly Media, Inc.
> 1005 Gravenstein Highway North
> Sebastopol, CA 95472
> 800-998-9938 (in the United States or Canada)
> 707-829-0515 (international or local)
> 707-829-0104 (fax)

We have a web page for this book, where we list errata, examples, and any additional information. You can access this page at *https://oreil.ly/AnalyticalSkills_AI_DS*.

Email *bookquestions@oreilly.com* to comment or ask technical questions about this book.

For news and more information about our books and courses, see our website at *http://www.oreilly.com*.

Find us on Facebook: *http://facebook.com/oreilly*

Follow us on Twitter: *http://twitter.com/oreillymedia*

Watch us on YouTube: *http://www.youtube.com/oreillymedia*

Acknowledgments

This book had three sources of inspiration. First, it has been the backbone in a Big Data for Managers course at the Tecnologico de Monterrey, in Mexico City. As such, I'm grateful to the university and the EGADE Business School specifically; they have provided a great place to think, discuss, and lecture on these ideas. Each cohort of students helped improve on the material, presentation, and use cases. To them I'm infinitely grateful.

My second source of inspiration came from my work as Head of Data Science at Telefonica Movistar Mexico and the wonderful team of data scientists that were there during my tenure. They helped create a highly energetic atmosphere where we could think out of the box and propose new projects to our business stakeholders.

I'm finally indebted to the different business people that I've encountered during my career, and especially during my tenure at Telefonica Movistar Mexico. It was never easy to sell these ideas, and the constant challenge helped improve my understanding of how they view the business, forcing me to build bridges between these two seemingly unrelated worlds.

I'm grateful to my family and friends for their support from the beginning. Finally, I'm infinitely grateful to my dogs, Matilda and Domingo. They were the perfect companions to the many long hours working on the book, always willing to cheer me up. We'll finally have more time to go to the park now.

Last but not least, I'm deeply grateful to my editor, Michele Cronin. Her suggestions dramatically helped improve the presentation of the book. I'd also like to thank Neal Ungerleider and Tom Fawcett for providing feedback on an early version of the book. I'm especially grateful to Katy Warr and Andreas Kaltenbrunner for providing very detailed commentary. This book is substantially better thanks to all of them. Needless to say, any mistakes that remain are my own.

## CHAPTER 1

Analytical Thinking and the AI-Driven Enterprise

It is April 2020, and the world is in the middle of a very serious global pandemic caused by novel coronavirus SARS-CoV-2 and the ensuing disease (COVID-19), with confirmed cases in the millions and deaths in the hundreds of thousands. Had you searched online for `AI coronavirus`, you could've found some very prestigious media and academic outlets highlighting the role that artificial intelligence (AI) can play in the battle against the pandemic (Figure 1-1).

What makes many uncomfortable with headlines like these is that they dress AI in a superhero suit that has become rather common, overstretching the limits of what can be achieved with AI today.

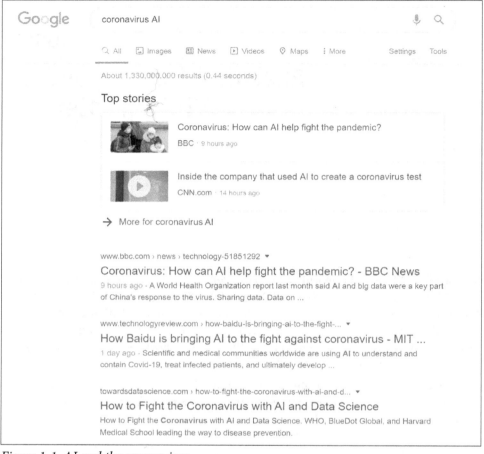

Figure 1-1. AI and the coronavirus

What Is AI?

If I had to divide the people of the world according to their understanding of the term "AI," I'd say there are four types of people.

On one end of the spectrum are those who've never heard the term. Since AI has become part of the popular folklore and is now a common theme in movies, TV shows, books, magazines, talk shows, and the like, I'd guess that this group is rather small.

Most people belong to a second group that believes that AI is closer to what practitioners call *Artificial General Intelligence* (AGI) or human-like intelligence. In their view, AI are humanoid-like machines that are able to complete the same tasks and make decisions like humans. For them, AI is no longer in the realm of science fiction as almost every day they come across some type of media coverage on how AI is changing our lives.

A third group, the practitioners, actually dislike the term and prefer to use the less sexy machine learning (ML) label to describe what they do. ML is mainly concerned with making accurate predictions with the use of powerful algorithms and vast amounts of data. There are many such algorithms, but the darling of ML techniques is known as deep learning—short for learning through deep neural networks—and is pretty much responsible for all the media attention the field gets nowadays.

To be sure, deep learning is also about using predictive algorithms that have proven quite powerful in tackling problems that a few years ago were only accessible to humans, specifically in the domains of image recognition and natural language processing (think Facebook automatically labelling your friends in a photo or virtual assistants like Alexa smoothing out your purchase experience on Amazon and controlling your lights and other devices connected to the internet at home).

I don't want to distract your attention with technical details, so if you want to learn more about these topics please consult the Appendix. The only thing I want to highlight here is that practitioners think "ML" when they hear or read "AI," and in their minds, this really just means *prediction algorithms*.

The fourth and final group is what I'll call "the experts," those very few individuals who are doing research, and are thus advancing the field of AI. These days most funds are directed toward pushing the boundaries in the field of deep learning, but in some cases they are doing significant research on other topics that aim at achieving AGI.

So what is AI? In this book I'll use AI and ML interchangeably since it has become the standard in the industry, but keep in mind that there are other topics besides prediction that are part of the AI research arena.

Why Current AI Won't Deliver on Its Promises

The trouble with AI starts with the name itself, as it inevitably makes us think about machines with human-like intelligence. But the difficulty comes not only from a misnomer but also from comments coming from within, as some recognized leaders in the field have reinforced expectations that will be hard to accomplish in the short term. One such leader claimed in 2016 that "pretty much anything that a normal person can do in <1 sec, we can now automate with AI" (*https://oreil.ly/IAFwY*). Others may be more cautious, but their firm conviction that deep neural networks are fundamental building blocks for achieving AGI provides the media with juicy headlines.

But I digress: what really matters for the purpose of this book is how this hype has affected the way we run our businesses. It is not uncommon to hear chief executive officers and other high-ranking executives say that they are disrupting their industries with AI. While they may not be fully aware of what the term entails, they are nonetheless backed by vendors and consultants that are very happy to share the riches before the bubble pops.

Hypes are risky because a natural response to unfulfilled expectations is to cut all funds and organizational focus.[1] My aim with this book is to show that while we may be far from creating human-like intelligence, with the current technology, we can create substantial value by transforming our companies into AI-driven enterprises. To do so we must start using AI as an input to improve our business decision-making capabilities.

Before that, let's understand how we got here, as this will help showcase some of the difficulties with the current approach and the opportunities that are already achievable.

How Did We Get Here?

Figure 1-2 shows the evolution of the top 10 global companies by market capitalization. With the exception of Berkshire Hathaway (Warren Buffett's conglomerate), Visa, and JPMorgan, all of the remaining companies are in the technology sector and all have embraced the data and AI revolutions.[2] At face value, this would suggest that if this worked for them, it must work for any other company. But is this the case?

---

1 The field of AI knows very well about this risk, as it has lived through at least two "winters" during which funding was almost entirely denied to any researcher.

2 Data from Wikipedia (*https://oreil.ly/p3fdX*), retrieved March 2020. In the plot, I use the information corresponding to the last quarter for each year only.

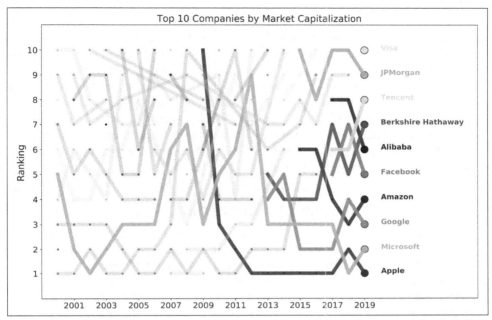

Figure 1-2. Evolution of market capitalization top-10 ranking—(companies that left the ranking before 2018 are not labeled)

Behind these successes, there are two stories that only converged recently. One has to do with the evolution of AI, and the other with the big data revolution.

The Data Revolution

Not so long ago, the queen of tech headlines was big data, and hardly anyone talked about AI (according to *The Economist*, in 2017 big data was the new oil (*https://oreil.ly/yePMT*)). Let's briefly tell the story of how big data rose to the crown and how AI surprisingly stole the spotlight in recent years.

In 2004, Google published its famous MapReduce paper (*https://oreil.ly/Dkd4x*) that enabled companies to distribute computation of large chunks of data (that wouldn't fit in a single computer) across different machines. Later, Yahoo! made its own open source version of the Google algorithm, marking the beginning of the data revolution.

It took a couple of years for technology commentators and consulting firms to start claiming that data would provide companies with endless opportunities for value creation. At the beginning, this revolution was built around one pillar: having more, diverse, and quickly accessible data. As the hype matured, two more pillars were added: predictive algorithms and a data-driven culture.

The three Vs

The first pillar involved the now well-known three Vs: *volume*, *variety*, and *velocity*. The internet transformation had provided companies with ever-increasing volumes of data. One 2018 estimate claimed that 90% of the data created in the history of humankind had been generated in the previous two years (*https://oreil.ly/aNciU*), and many such calculations abound. Technology had to adapt if we wanted to analyze this apparently unlimited supply of information. We not only had to store and process larger amounts of data, but we also needed to deal with new unstructured types of data such as text, images, videos, and recordings that were not easily stored or processed with the data infrastructure available at the time.

Structured and Unstructured Data

The second V, *variety*, emphasizes the importance of analyzing all types of data, not just structured data. If you have never heard of this distinction, think of your favorite spreadsheet program (Excel, Google Sheets, etc.). These programs organize information in tabular arrangements of rows and columns that provide a lot of structure so that we can efficiently process information within a user-friendly interface. This is a simple example of structured data: anything you can store and *analyze* using rows and columns belongs to this class.

Have you ever copied and pasted an image in Excel? Not only can you paste images, but you can also use it to store entire texts and even videos. But the fact that you can paste them doesn't mean you can *analyze* them. And storage isn't efficient either: you can save a lot of space on disk by using some type of compression or efficient formats. *Unstructured* datasets are not efficiently stored or analyzed using tabular formats, and these include all types of multimedia (images, videos, tweets, etc.). Now, these provide *a lot* of valuable information for companies, so why shouldn't we use them?

After the innovations were made, consultants and vendors came up with new ways to market these new technologies. Before the age of big data, the Enterprise Data Warehouse was used to store and analyze structured data. The new age needed something equally new, and thus the *data lake* was born with the promise of providing flexibility and computational power to store and analyze big data.

Thanks to "linear scalability," if twice the work needed to be done, we would just have to install twice the computing power to meet the same deadlines. Similarly, for a given task, we could cut the current time in half by doubling the amount of infrastructure. Computing power could be easily added by way of commodity hardware, efficiently operated by open source software readily available for us to use. But the data lake also allowed for quick access to the larger variety of data sources.

Once we tackled the volume and variety problems, velocity was the next frontier, and our objective had to be the reduction of time-to-action and time-to-decision. We were now able to store and process large amounts of very diverse data in real time or near-real time if necessary. The three Vs were readily achievable for any company willing to invest in the technology and the know-how. Nonetheless, the riches were not in sight yet, so two new pillars were added—prediction and data-driven culture— along with a recipe for success.

Data maturity models

Since data alone was not creating the value that was promised, we needed some extra guidance; this is where maturity models entered with the promise of helping companies navigate through the turbulent waters created by the data revolution. One such model is depicted in Figure 1-3, which I will explain now.

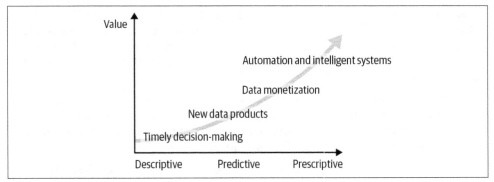

Figure 1-3. A possible data maturity model showing a hierarchy of value creation

Descriptive stage. Starting from the left, one thing was apparent from the outset: having more, better, and timely data could provide a more granular view of our businesses' performance. And our ability to react quickly would certainly allow us to create some value. A health analogy may help to understand why.

Imagine you install sensors in your body, either externally through wearables or by means of other soon-to-be-invented internal devices, that provide you with more, better, and timely data on your health. Since you may now know when your heart rate or your blood pressure increases above some critical level, you can take whatever measures are needed to bring things back to normal. Similarly, you can track your sleeping patterns or sugar levels and adjust your daily habits accordingly. If we react fast enough, this newly available data may even save our lives. This kind of descriptive analysis of past data may provide some insights about your health, and the creation of value depends critically on our ability to react quickly enough.

Predictive stage. But more often than not it's too late when we react. Can we do better? One approach would be to replace reaction with predictive action. As long as predictive power is strong enough, this layer should buy us time to find better actions, and thus, new opportunities to create value.

This new stage allowed us to develop new *data products*, such as recommendation engines (think Netflix), and it also gave rise to the age of data monetization. The online advertising business was thus born, marking an important inflection point in our story. The dream of marketers came to life with the promise of *selling the right product to the right person at the right time*, all thanks to data and the predictions created with it.

Importance of Online Advertising

Most of the riches created by big data were the product of the success of online advertising. The online advertising business is huge and highly lucrative. One source estimates that more than $500 billion will be spent during 2023 across the globe (*https://oreil.ly/8dJ8a*). If that figure alone doesn't say much, consider that it is close to Belgium's Gross Domestic Product (*https://oreil.ly/K7BI7*).

The two main players in this business are Google and Facebook. They have built their businesses largely funded by the revenues from this profitable industry, and thanks to the riches that came with it, they have been able to fund the fast recent development in the AI arena (many times through acquisitions).

So it seems fair to say that the success of big data in online advertising has played an important role in facilitating AI's current boom.

Prescriptive stage. The top rank in this hierarchy of value creation is taken by our ability to automate and design intelligent systems. We are now at the *prescriptive* layer: once you have enough predictive power you can start finding the *best* actions for your business objectives. This is the layer where firms move from prediction to optimization, the throne in the data Olympus, and interestingly enough, this is the least explored step in most maturity models.

A Tale of Unrealized Expectations

In less than 15 years, we've lived through two booms—the big data revolution followed by the current AI stage—so you may wonder why the promises have yet to be fulfilled.

I'm not a big fan of data maturity models, but I believe the answer lies within them: *most companies have yet to arrive at the prescriptive stage*. Big data was all about the descriptive stage, and as we've mentioned, AI is primarily concerned with prediction. Since everything has been laid out for us in the past few years, the question about what's behind our apparent inability to move forward remains.

I'm convinced that market forces are an important factor, meaning that once a hype begins, market players want to reap the benefits until completely exhausted before moving on to the next big thing. Since we're still in that phase, there are no incentives to move forward yet.

But it is also true that to become prescriptive we need to acquire a new set of analytical skills. As of today, with the current technology, this stage is done by humans, so we need to prepare humans to pose and solve prescriptive problems. This book aims at taking us closer to that objective.

Analytical Skills for the Modern AI-Driven Enterprise

Tom Davenport's now classic *Competing on Analytics* (Harvard Business Press) pretty much equates analytical thinking with what later came to be known as data-drivenness: "By analytics we mean the extensive use of data, statistical and quantitative analysis, explanatory and predictive models, and fact-based management to drive decisions and actions." One alternative definition can be found in Albert Rutherford's *The Analytical Mind* (independently published): "Analytical skills are, simply put, problem-solving skills. They are characteristics and abilities that allow you to approach problems in a logical, rational manner in an effort to sort out the best solution."

In this book I will define *analytical reasoning* as the ability to translate business problems into *prescriptive solutions*. This ability entails both being data-driven and being able to solve problems rationally and logically, so the definition is in fact in accordance with the two described previously.

To make things practical, I will equate business *problems* with business *decisions*. Other problems that are purely informative and do not entail actions may have intrinsic value for some companies, but I will not treat them here, as my interest is in creating value through analytical *decision-making*. Since most decisions are made without knowing the actual consequences, AI will be our weapon to embrace this intrinsic uncertainty. Notice that under this approach, prediction technologies are important *inputs* into our decision-making process but not the *end*. Improvements in the quality of predictions can have first- or second-order effects depending on whether we are already making near-to-optimal choices.

Key Takeways

- *Most companies haven't been able to create value through data or AI in a sustainable and systematic way*: nonetheless, many have already embarked on their own efforts just to reach a wall of disappointment.

- *Today's AI is about prediction*: AI is overhyped, not only because of its deceiving name, but also because there is only so much one can achieve through better prediction. These days, AI most commonly refers to deep learning. Deep neural networks are highly nonlinear prediction algorithms that have shown remarkable success in the areas of image recognition and natural language processing.

- *Before AI, we had the big data revolution*: the data revolution preceded the current hype and also came with the promise to generate outstanding business results. It was built around the three Vs—volume, variety, and velocity—and later complemented with prediction algorithms and data-driven culture.

- *Data and prediction cannot create sustainable value by themselves*: maturity models suggest that value is created by making optimal decisions in a data-driven way. For this, we need data and prediction as inputs in our decision-making process.

- *We need a new set of analytical skills to be successful in this prescriptive stage*: current technology precludes us from automating the process of translating business problems into prescriptive solutions. Since humans need to be involved all along the way, we need to upscale our skillset to capture all the value from data- and AI-driven decision-making.

Further Reading

2019 and 2020 witnessed a very interesting debate on the limits of what can be achieved through AI. You can see one such debate in the discussion that Gary Marcus and Joshua Bengio had in Montreal (*https://oreil.ly/MSCrc*). If you prefer reading, Gary Marcus and Ernst Davis's *Rebooting AI: Building Artificial Intelligence We Can Trust* (Pantheon) will provide many of the details on why many are critical about deep learning being the way to achieve AGI.

On the topic of how AI will affect businesses, I highly recommend *Prediction Machines: The Simple Economics of Artificial Intelligence* by Ajay Agrawal, Joshua Gans, and Avi Goldfarb (Harvard Business Press). Written by three economists and AI strategists, the book provides a much-needed, away-from-the-hype, down-to-earth account of current AI. Their key takeaway is that thanks to current developments, the cost of predictive solutions within the firm has fallen considerably while quality has kept increasing, providing great opportunities for companies to transform their business models. Also written by economists, *Machine Platform Crowd: Harnessing Our Digital Future* by Andrew McAfee and Erik Brynjolfsson (W. W. Norton and Company) discusses how the data, artificial intelligence, and digital transformations are affecting our businesses, the economy, and society as a whole.

Data maturity models appear in several books: you can check out Thomas Davenport and Jeane Harris's *Competing on Analytics* (Harvard Business Press); *Big Data at Work: Dispelling the Myths, Uncovering the Opportunities* also by Tom Davenport (Harvard Business Press); or Bill Schmarzo's *Big Data: Understanding How Data Powers Big Business* (Wiley).

If you're interested in learning more about our quest to achieve AGI, Nick Bostrom's *Superintelligence. Paths, Dangers, Strategies* (Oxford Univeristy Press) discusses at great length and depth what intelligence is and how superintelligence could emerge, as well as the dangers from this development and how it can affect society. Similar discussions can be found in Max Tegmark's *Life 3.0. Being Human in the Age of Artificial Intelligence* (Vintage).

Finally, on the podcast side, I recommend following Lex Fridman's *Artificial Intelligence* (*https://lexfridman.com/ai*). There are many great interviews with leaders in the field that will provide much more context on the current state of affairs.

## CHAPTER 2

Intro to Analytical Thinking

In the last chapter, I defined *analytical thinking* as the ability to translate business problems into prescriptive solutions. There is a lot to unpack from this definition, and this will be our task in this chapter.

To really understand the power of prescriptive solutions, I will start by precisely defining each of the three stages present in any analysis of business decisions: these are the descriptive, predictive, and prescriptive steps we have already mentioned in Chapter 1.

Since one crucial skill in our analytical toolbox will be formulating the right business questions from the outset, I will provide an initial glimpse into this topic. Spoiler alert: we only care about business questions that entail business decisions. We will then dissect decisions into levers, consequences, and business results. The link between levers and consequences is intermediated by *causation*, so I will spend quite a bit of time talking about this topic. Finally, I will talk about the role that uncertainty plays in business decisions. Each of these topics is tied to one skill that will be developed throughout the book.

What Is a Lever?

In the context of this book, "levers" are synonymous with "actions" or "decisions," so whenever we say that "we want to pull some lever to obtain a business outcome," this means that we are looking for suitable actions or decisions.

Descriptive, Predictive, and Prescriptive Questions

In Chapter 1, we saw that data maturity models usually depict a nice, smooth road that starts at the descriptive stage, goes through the predictive plateau, and finally ascends to the predictive summit. But why is this the case? Let's start by understanding what these mean, and then we can discuss why commentators and practitioners alike believe that this is the natural ascension of the data evolution.

In a nutshell, *descriptive* relates to how things are, *predictive* to how we believe things will be, and *prescriptive* to how things ought to be. Take Tyrion Lannister's quote in the *Game of Thrones* "The Dance of Dragons" episode: "It's easy to confuse what is with what *ought* to be, especially when *what is* has worked out in your favor" (my emphasis). Tyrion seems to be claiming that we have the tendency to confuse the descriptive and prescriptive when things turn out well, in what may well be a form of confirmation bias. Incidentally, when the outcome is negative, our tendency is to think that this was the worst possible result and attribute our fate to some version of Murphy's Law.

In any case, as this discussion shows, the prescriptive stage is a place where we can rank different options so that words like "best" or "worst" make any sense at all. It follows that the prescriptive layer can never be inferior to the descriptive one, as in the former we can always make the best decision.

But what about prediction? To start, its intermediate ranking is at least problematic, since description relates to the current state and prescription to the *quality of decisions*, and prediction is an input to make decisions, which may or may not be optimal or even good. The implicit assumption in all maturity models is that the quality of decisions can be improved when we have better predictions about the underlying uncertainty in the problem; that good predictions allow us to plan ahead and move proactively, instead of reacting to the past with little or no room to maneuver. That said, this really is an assumption as there's nothing inherent about prediction that makes it improve the outcomes for our businesses.

When Predictive Analysis Is Powerful: The Case of Cancer Detection

Let's take an example where better prediction can make a huge difference: cancer detection (*https://oreil.ly/00Otb*). Oncologists usually use some type of visual aid such as X-rays or the more advanced CT scans for early detection of different pathologies. In the case of lung cancer, an X-ray or a CT scan is a description of the patient's current health status. Unfortunately, visual inspection is ineffective unless the disease has already reached a late stage, so description here, by itself, may not provide enough time for a proactive reaction. AI has shown remarkable prowess in predicting the existence of lung cancer from inspecting CT scans (*https://oreil.ly/8ZU6D*), by identifying spots that will eventually turn out to be malignant. But prediction can only take us so far. A doctor should then recommend the right course of action for the patient to fully recover. AI provides the predictive muscle, but humans prescribe the treatment.

Descriptive Analysis: The Case of Customer Churn

Let's run a somewhat typical descriptive analysis of a use case that most companies have dealt with: customer churn or attrition. We will see that without guidance from our business objectives, this type of analysis might take us to a dead end.

What Is Customer Churn?

In case you don't recognize the term, customer churn is the rate at which customers stop using a company's product or service per period of time. For instance, if your company's monthly rate of churn is 5%, this means that 5 out of 100 customers that were purchasing from you at the beginning of the period are no longer doing so when the month ends. As you might imagine, the exact definition varies from industry to industry and depends crucially on the expected frequency of purchases (think about a credit card).

The main reason we care about churn is that our customer acquisition costs are generally substantially larger than the corresponding retention costs, so having a proactive churn control strategy has become an objective in and of itself.

Describing churn

Suppose that your boss wants to get churn under control. As a first step, she may ask you to diagnose the magnitude of the problem. After wrangling with the data, you come up with the following two plots (Figure 2-1). The left plot shows a time series of daily churn rates. Confidently, you state two things: after having a relatively stable beginning of the year, churn is now on the rise. Second, there is a clear seasonal pattern, with weekends having lower than average churn. In the right panel you show that municipalities with higher average incomes also have higher churn rates, which of course is a cause for concern since your most valuable customers may be switching to other companies.

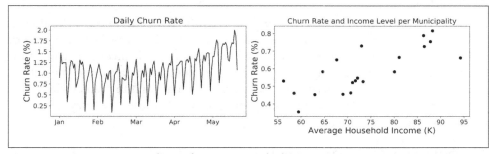

Figure 2-1. Descriptive analysis of our company's churn rate

This is a great example of what can be achieved with descriptive analysis, thanks to our remarkable ability to recognize patterns in the data. Here we quickly identified a change in the trend (churn is accelerating), the existence of strong seasonal effects, and a positive correlation between churn rates and average household income in the scatterplot.

But this also highlights some of its shortcomings. First, as you've probably heard, *correlation does not imply causation*, a topic that will be discussed at length later in this chapter. Related to this, successful root cause analysis requires our ability to create theories about cause and effect. Without these theories we cannot aim at providing alternative courses of action to improve our company's situation. Inspecting data without advancing some plausible explanations is the perfect recipe for making your analytics and data science teams waste valuable time.

The Trap of Finding Actionable Insights

One common catchphrase among consultants and vendors of big data solutions is that once they are given enough data, your data analysts and data scientists will be able to find *actionable insights*.

This is a common trap among business people and novice data practitioners: the idea that given some data, if we inspect it long enough, these actionable insights will emerge, almost magically. I've seen teams spend weeks waiting for the actionable insights to appear, without any luck.

Experienced practitioners reverse engineer the problem: start with the question, formulate hypotheses, and use your descriptive analysis to find evidence against or in favor of these hypotheses. Note the difference: under this approach we actively search for actionable insights by first deciding where to look for them, as opposed to waiting for them to emerge from chaos.

Predicting churn

As a next step, your boss may ask you to *predict* churn in the future. How should you proceed? It really depends on what you want to achieve with this analysis. If you work in finance, for example, and you're interested in forecasting the income statement for the next quarter, you'd be happy to predict aggregate churn rates into the future. If you are in the marketing department, however, you may want to predict which customers are at risk of leaving the company, possibly because you may try using different retention campaigns.

Prescribing courses of action to reduce churn

Finally, suppose that your boss asks you to recommend alternative courses of action to *reduce* the rate of customer churn. This is where the prescriptive toolkit becomes

---

quite handy and where the impact of making good decisions can be most appreciated. You may then pose a cost-benefit analysis for customer retention and come up with a rule that maximizes your customer lifetime value (CLV).

Customer Lifetime Value (CLV)

How should we value our customers? One approach is to assign the current value derived from each one of them. The problem with this short-term view is that companies invest in their customers all the time, from acquisition to retention, marketing, etc., so to value those investments we also need the long-run view from the revenues side.

Several decades ago, people started looking at customers as assets (*https://oreil.ly/ gP0kd*), and under this approach, the right metric is the *stream* of profits derived from them. One difficulty with the stream approach is that at any time our customers may decide to change companies, so we need to incorporate an uncertain time window into the analysis.

The CLV measures the discounted present value of all profits obtained from a relationship with one customer along their expected duration with the company.

For instance, assuming a monthly discount rate of 1%, a new customer who will keep purchasing our goods and services for the next 11 months, leaving a monthly profit of 1 dollar, will have a CLV of $1 + \$1/(1.01) + \$1/(1.01)^2 + \cdots + \$1/(1.01)^{10} = 10.4$ dollars. In practice, to compute the CLV we need an estimate of the expected duration of a customer's relationship with us, as well as an estimate of how profits change over time.

We will have the opportunity to go into greater detail on this use case, but let me just single out two characteristics of any prescriptive analysis: as opposed to the two previous analyses, here we actively recommend courses of action that can improve our position, by way of incentivizing a likely-to-leave customer to stay longer with us. Second, prediction is used as an input in the decision-making process, helping us calculate *expected* savings and costs. AI will help us better estimate these quantities, which is necessary for our proposed decision rule. But it is this decision rule that creates value, not prediction itself.

One of the objectives of this book is to prepare us to translate business questions into prescriptive solutions, so don't worry if it's not obvious yet. We will have time to go through many step-by-step examples.

Business Questions and KPIs

One foundational idea in the book is that value is derived from *making decisions*. As such, prediction in the form of machine learning is just an input to create value. In this book, whenever we talk about business questions, we will always have in mind business decisions. Surely, there are business questions that are purely informative and no actions are involved. But since our aim is to systematically create value, we will only consider actionable questions. As a matter of fact, one byproduct of this book is that we will learn to look for actionable insights in an almost automatic fashion.

It suggests the question, then, of *why* we have to make a decision. Only by answering this question will we be able to know how to measure the appropriateness or not of the choices we make. Decisions that cannot be judged in the face of any relevant evidence are to be discarded. As such, we will have to learn how to select the right metrics to track our performance. Many data science projects and business decisions fail not because of the logic used but because the metrics were not right for the problem.

There is a whole literature on how to select the right key performance indicators (KPIs), and I believe I have little to add on this topic. The two main characteristics I look for are *relevance* and *measurability*. A KPI is relevant when it allows us to clearly assess the results from our decisions *with respect* to the business objective. Notice that this doesn't have to do with how pertinent the business question is, but rather, with whether we are able to evaluate if the decision worked or not, and by how much. It follows that a good KPI should be measurable, and this should be with little or no delay with respect to the time when the decision was made. Not only is there an opportunity cost of delayed measurement, but it may also be harder to identify the root cause.

KPIs to Measure the Success of a Loyalty Program

Let's briefly discuss one example. Suppose that our chief marketing officer asks us to evaluate the creation of a loyalty program for the company. Since the question starts with an action (i.e., to create the loyalty program or not), it immediately registers for us as a business problem. What metrics should we track? To answer this let's start the sequence of *why* questions.

The Sequence of Why Questions

The following example showcases a technique that I call *the sequence of* why *questions*. It is used to identify the business metric that we want to optimize.

It works by starting with what you, your boss, or your colleagues may think you want to achieve and questions the reasons for focusing on this objective. Move one step above and repeat. It terminates when you're satisfied with the answer. Just in passing, recall that to be satisfied you must have a relevant and measurable KPI to quantify the business outcome you will focus on.

Our *why* questions, then, are as follows:

- Create a loyalty program. *Why?*
- Because you want to reward loyal customers. *Why?*
- Because you want to incentivize customers to stay longer with the company. *Why?*
- Because you want to increase your revenues in the longer term. *Why?*

And of course, the list can go on. The important thing is that the final answer to these questions will usually let you clearly identify what KPI is relevant for the problem at hand, and any intermediate metrics that may provide useful; if it's also measurable, then you have found the right metric for your problem.

Consider the second question, for example. Why would anyone want to reward loyal customers? They are already loyal, without the need for any extrinsic motivation, so this strategy may even backfire. But putting aside the underlying reasoning, why is loyalty meaningful and how would you go about measuring the impact of the reward? I argue that loyalty by itself is not meaningful: we prefer loyal customers to not-so-loyal customers because they represent a more stable stream of revenues in the future. If you're not convinced, think about those loyal but *unprofitable* customers. Do you still rank their loyalty as high as before? If loyalty per se is not what you're pursuing, then you should keep going down the sequence of *why* questions.

Just for the sake of the discussion, suppose that you still want to reward loyal customers. How do we measure if the program worked, or put differently, what is a good KPI for this? One commonly used method is to directly ask our customers, as done with the Net Promoter Score (NPS). To calculate the NPS we first ask our customers how likely they are to recommend us as a company on a scale from 0 to 10. We then classify them into *Promoters* (9 to 10), *Detractors* (0 to 6), and *Passive* (7, 8). Individual answers are finally aggregated into the NPS by subtracting the percentage of detractors from the percentage of promoters.

On the bright side, this is a pretty *direct* assessment: we just go and ask our customers if they value the reward. It can't get more straightforward than that. The problem here is that humans act on motivations, so we generally can't tell if the answer is truthful, or if there is some underlying motive and they're trying to game our system. This type of strategic consideration matters when we assess the impact of our decisions.

An alternative is to let the customers indirectly *reveal* their level of satisfaction through their actions, say from the amount or frequency of their recent transactions, or through a lower churn rate for those who receive the reward relative to a well-designed control group.[1] Companies will always have customer surveys, and they should be treated as a potentially rich source of information. But a good practice is to always check if what they *say* is supported by their actions.

An Anatomy of a Decision: A Simple Decomposition

Figure 2-2 shows the general framework we will use to decompose and understand business decisions. Starting from the right, it is useful to repeat one more time that we *always start with the business*. If your objective is unclear or fuzzy, most likely the decision shouldn't be made at all. Companies tend to have a bias for action, so fruit-less decisions are sometimes made. This may not only have unintended negative consequences on the business side; it could also take a toll on employees' energy and morale. Moreover, we now take for granted that our business objective can be measured through relevant KPIs. This is not to say that metrics arise naturally: as highlighted in a later example, we must choose our metrics carefully.

It is generally the case that we can't simply manipulate those business objectives ourselves (remember Enron? (*https://oreil.ly/Dh_H4*)), so we need to take some actions or pull some levers in order to try to generate results. Actions themselves map to a set of consequences that directly affect our business objective. To be sure: *we* pull the levers, and our business objectives depend on consequences that arise when the "environment" reacts. The environment can be humans or technology, as we will see later.

---

1 We will talk about designing experiments or A/B tests later in this chapter.

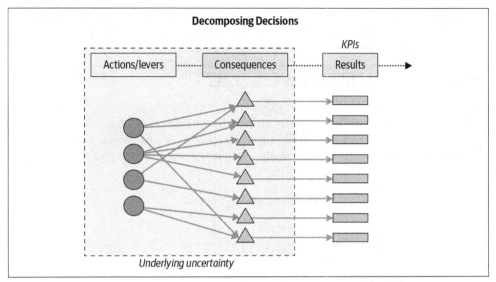

Figure 2-2. Decomposing decisions: actions, consequences, and business results

Even if the mapping is straightforward (most times it isn't), it's still mediated by uncertainty, since at the time of the decision it is impossible to know exactly what the consequences will be. We will use the powers of AI to embrace this underlying uncertainty, allowing us to make better decisions. But make no mistake: *value is derived from the decision, and prediction is an input to make better decisions.*

Difference Between Actions, Consequences, and Results

In case you haven't figured out the role that consequences play in the decomposition, here's an example. Suppose that our objective is to increase our revenues. To do so we decided to pull the pricing lever and offer some discounts to our customers. The consequence from our action is that our customers increase their spend on our brand, which itself generates higher revenues.

- *Action*: offer a discount
- *Consequence*: customers increase their demand for our product
- *Outcome*: revenues increase

To sum up, in our daily lives and in business, we generally pursue well-chosen, measurable objectives. Decision-making is the act of choosing among competing actions to attain these objectives. Data-driven decision-making is acting upon evidence to assess alternative courses of action. Prescriptive decision-making is the science of choosing the action that produces the best results for us; we must therefore be able to rank our choices relative to a measurable and relevant KPI.

An Example: Why Did You Buy This Book?

One example should illustrate how this decomposition works for *every* decision we make (Figure 2-3). Take your choice to purchase this book. This is an action you already made, but, surely, you could have decided otherwise. Since we always start with the business problem, let me imagine what type of problem you were trying to solve.

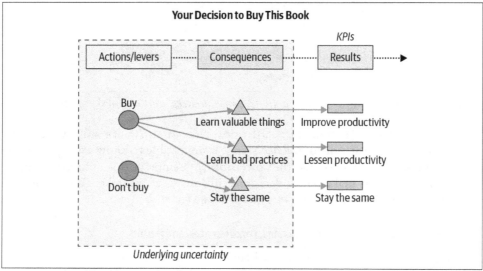

Figure 2-3. Decomposing your decision to buy this book

I don't know what objective you were solving when you decided to make the purchase, but in my case, I would've been interested in advancing my career. I will thus assume that the key metric you want to optimize is your productivity, and following our discussion on KPIs, I will conveniently assume that it is measurable.

Since you're reading the book, I'll also simplify all interesting details and just take two possible actions: buy or not buy. As the figure shows, if you buy the book there are at least three possible consequences: you learn valuable things, you learn bad practices, or you learn nothing. Naturally, each of these consequences impact your productivity.

If you don't buy the book many things can happen. For instance, you may get a sudden burst of inspiration and start understanding the intricacies of your job, thereby improving your productivity. Though plausible, we will appeal to Occam's razor and keep the most likely consequence that your knowledge and productivity stay the same.

 Occam's Razor

When there are many plausible explanations for a problem, the principle known as Occam's razor appeals for the simplest one. Similarly, in statistics, when we have many possible models to explain an outcome, if we apply this principle we would attempt to use the most parsimonious one.

Don't worry if this isn't entirely clear now; Chapter 5 will be devoted entirely to improving our simplifying skills.

Finally, the difficulty here is that you don't really know what consequence will follow at the time you make the decision. For instance, contrary to your beliefs, it could be that O'Reilly made a mistake by signing this book or author. Unfortunately you will only know once you read it (so please do). This is the underlying uncertainty of this specific decision.

To sum up, notice how a simple action helped us to clearly and logically find the problem being solved, a set of levers, their consequences, and the underlying uncertainty. You can use this decomposition with any decision you make.

A Primer on Causation

The upcoming chapters will delve into each of the stages in the decomposition, so there will be enough time to understand where these levers come from and how they map to consequences. It is important, though, to stop now and recognize that this mapping is mediated by *causal* forces.

Going back to the saying that "correlation does not imply causation," no matter how many times we've heard about it, it is still very common to get the two terms confused. Our brain evolved to become a powerful pattern-recognizing machine, but we are not so well equipped to distinguish causation from correlation.[2]

Defining Correlation and Causation

Strictly speaking, correlation is the presence or absence of any linear dependencies in two or more variables. Less formally, two variables are correlated if they tend to "move together."

---

2 To be fair, even after taking into account this apparent impairment, we are by far the most sophisticated causal creatures that we know of, and we are infinitely superior to machines (since at the time of writing, they completely lack the ability, and it is not even clear when this ability may be achieved or if it's achievable at all).

Causality is harder to define, so let us take the shortcut followed by almost everyone: a relation of causality is one of cause and effect. X (partially) causes Y if Y is (partially) an effect of X. The "partial" qualifier is used because rarely is one factor the unique source of a relationship.

One can also define causality in terms of *counterfactuals*: *had X not taken place, is it true that Y had been observed?* If the answer is positive, then it is unlikely that a causal relationship from X to Y exists. Again, the qualifier "unlikely" is important and related to the previous "partial" qualifier: there are causal relations that only occur if the right combination of conditions is present.

Scatterplots like the one in Figure 2-1 are very good at depicting correlations between two variables, but unfortunately can't guide us in our quest to understand causation. To do so, it's quite standard to ask counterfactual questions in both directions and use Occam's razor to select a subset of plausible explanations.

Some Difficulties in Estimating Causal Effects

Estimating the causal impact on outcome Y of pulling a lever $X \Longrightarrow Y$ is paramount since we are trying to engineer optimal decision-making. The analogy is not an accident: like the engineer who has to understand the laws of physics to build skyscrapers, bridges, cars, or planes, the analytical leaders of today must have some level of understanding of the causal laws mediating our own actions and their consequences to make the best possible decisions. And this is something that humans must do; AI will help us later in the decision-making process, but we must first overcome the causal hurdles.

Problem 1: We can't observe counterfactuals

As discussed in the previous sections, there are several problems that make our identification of causal effects much harder. The first one is that we only observe the facts, so we must imagine alternative *counterfactual* scenarios. It is an understatement that one of the most important skills analytical thinkers must develop is to question the initial interpretation given to empirical results, and to come up with counterfactual alternatives to be tested. Would the consequences be different had we pulled different levers, or the same levers but under different conditions?

Let's stop briefly to discuss what this question entails. Suppose we want to increase lead conversion in our telemarketing campaigns. Tom, a junior analyst who took one class in college on Freudian psychoanalysis, suggests that female call center representatives should have higher conversion rates, so the company decides to have its very capable group of female representatives make all outbound calls for a day. The next day, they meet to review the results: lead conversion went from the normal 5% to an outstanding 8.3%. It appears that Freud was right, or better, that Tom's decision to take the class had finally proven correct. Or does it?

To get the right answer, we need to imagine a customer receiving one call from the female representative in one universe, and the *exact same* call from a male representative in a parallel universe (Figure 2-4). Exact customer, exact timing, exact mood, and exact message; everything is the same in the two scenarios: we only change the tone of voice from that of a male to a female. Needless to say, putting in practice such a counterfactual sounds impossible. Later in this chapter, we will describe how we can simulate these impossible counterfactuals through well-designed randomized experiments or A/B tests.

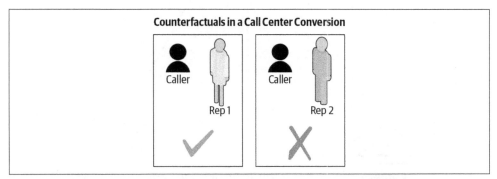

Figure 2-4. Counterfactual analysis of lead conversion rates in a call center

Problem 2: Heterogeneity

A second problem is *heterogeneity*. Humans are intrinsically different, each and every one the product of both our genetic makeup and lifetime experiences, creating unique worldviews and behaviors. Our task is not only to estimate how behavior changes when we choose to pull a specific lever—the causal effect—but we must also take care of the fact that different customers react differently. An influencer recommending our product will have different effects on you and me: I may now be willing to try it, while you may choose to remain loyal to your favorite brand. How do we even measure heterogenous effects?

Figure 2-5 shows the famous bell curve, the normal distribution, the darling of statistical aficionados. I'm using it here to represent the natural variation we may encounter when analyzing our customers' response when our influencer recommends our product. Some of his followers, like me, will accept the cue and react positively—represented as an action right of the vertical dashed line, the average response across all followers, followers' followers, and so on. Some will have no reaction whatsoever, and some may even react negatively—that's the beauty of human behavior; we sometimes get the full spectrum of possible actions and reactions. The shape of the distribution has important implications, and in reality, our responses may not be as symmetric; we may have longer left or right tails and reactions may be skewed toward the positive or the negative. The important thing here is that people react differently, making things even more difficult for us when we try to estimate a causal effect.

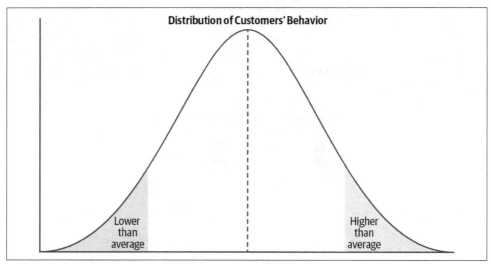

Figure 2-5. A normal distribution as a way to think about customer heterogeneity

The way we usually deal with heterogeneity is by dispensing of it by estimating a unique response, usually given by the average or the mean (the vertical line in Figure 2-5). The mean, however, is overly sensitive to extreme observations, so we may sometimes replace it with the median, which has the property that 50% of responses are lower (to the left) and 50% higher (to the right); with bell-shaped distributions the mean and the median are conveniently the same.

Problem 3: Confounders

When searching for causal relationships it's quite common to start by plotting scatterplots like the one in Figure 2-6 where each marker denotes a pair of (x, y) observations.

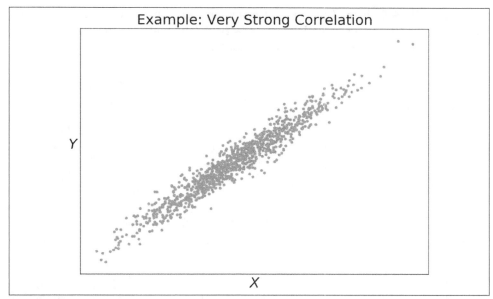

Figure 2-6. A simulation of two highly correlated variables

You may be tempted to assert that in this case there is clear evidence that *X* causes *Y* or vice versa—it is common to interpret scatterplots as relationships from the variable in the horizontal axis to outcomes on the vertical axis—but as Example 2-1 shows, this interpretation is faulty:

Example 2-1. Simulating the effect of a third unaccounted variable on the correlation of the other two

```
# fix a seed for our random number generator and number of observations to simulate
np.random.seed(422019)
nobs = 1000
# our third variable will be standard normal
z = np.random.randn(nobs,1)
# let's say that z --> x and z--> y
# Notice that x and y are not related!
x = 0.5 + 0.4*z + 0.1*np.random.randn(nobs,1)
y = 1.5 + 0.2*z + 0.01*np.random.randn(nobs,1)
```

To be sure, a third variable *z* positively affects both *x* and *y*, creating this spurious correlation. If we can control for this third variable (also known as a *confounder*), we may be able to get a better sense of the net relationship between the two variables of interest.

Consider the examples shown in Figure 2-7. The top left panel plots a measure of global CO_2 emissions and per capita real Gross Domestic Product (GDP) in Mexico for the period 1900–2016. The top right panel plots the number of divorces in Wales and England against Mexican GDP for 1900–2014. The bottom panel plots the three time series, indexed so that the 1900 observation is 100.[3]

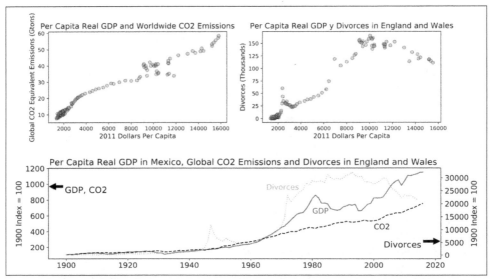

Figure 2-7. The top left panel plots global CO$_2$ emissions against real per capita Gross Domestic Product (GDP) for Mexico for the period 1900–2016; the top right panel does the same, replacing CO$_2$ emission with the number of divorces in Wales and England during 1900–2014; the bottom plot shows the time series for each of these variables

If we just inspected the scatterplots, we would be tempted to conclude that global emissions and divorces in the UK are somehow causally related to economic growth in Mexico. In this case, however, a third variable is responsible for such spurious correlation: statisticians and econometricians call a *time trend* the natural growth rate of a variable when plotted against time. The bottom panel shows that indeed these growth rates were very similar across the three variables in specific time periods.

Once we identify a confounder we can just *control* for it in our predictive algorithms (see the Appendix). But the problem of finding confounders is far from straightforward, so this task has to be done by us (and is thus not easily automatable).

3 Sources: GDP data comes from *https://oreil.ly/9J_wb*. CO$_2$ emissions from *https://oreil.ly/9J3XF*. Divorce rates from *https://oreil.ly/t_1x-*.

Problem 4: Selection effects

One final problem is the prevalence of selection effects. This usually arises because we choose the customer segments we want to act upon, or customers self-select themselves, or both. An important result in causal inference is that if we wish to estimate the causal effect from a treatment by comparing the average outcomes of two groups, we need to find a way to eliminate selection bias.[4]

Selection Bias and Causal Effects

Because of selection bias we may over- or underestimate a causal effect when we just take the difference in average outcomes across treated and control groups. Stated as an equation:

Observed Difference in Means = Causal Effect + Selection Bias

It is standard practice to plot average outcomes as in the top panel of Figure 2-8. In this case, the outcome for the control is 0.29 units (let's say hundreds of dollars) higher than for those exposed to our action or lever. This number corresponds to the left-hand side of the previous equation. The bottom panel shows the corresponding distributions of outcomes. Using the mean to calculate differences is standard practice, but it is useful to remember that there are a full spectrum of responses, in some cases with a clear overlap between the two groups: the shaded areas show responses from customers in the two groups that are indistinguishable from each other.

---

4 Hereafter I will use the term "treated" or "those who receive a treatment" to refer to those customers that are exposed to our action or lever. This jargon is common in the statistical analysis of experiments and was originally borrowed from the analysis of medical trials.

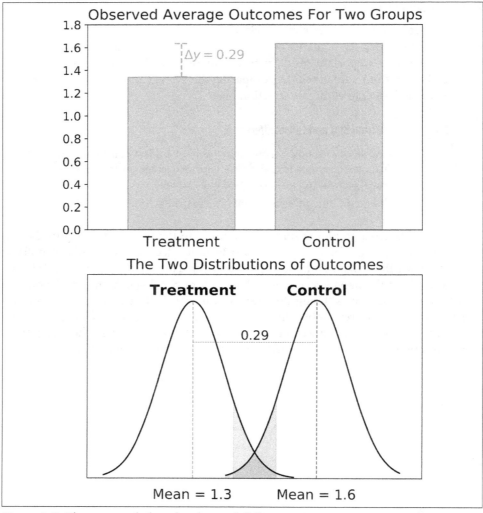

Figure 2-8. The top panel plots the observed differences in average outcomes for treatment and control groups; the bottom panel shows the actual distributions of outcomes

In any case, the difference in observed outcomes (left-hand side) is not enough for us since we already know that it is potentially biased by selection effects; since our interest is in estimating the causal effect, we must therefore devise a method to cancel this pervasive effect.

Statisticians and econometricians, not to mention philosophers and scientists, have been thinking about this problem for centuries. Since it is physically impossible to get an exact copy of each of our customers, is there a way to assign our treatments and circumvent the selection bias? It was Ronald A. Fisher, the famous 20th-century statistician and scientist, who put on firm grounds the method of experimentation, the most prevalent among practitioners when we want to estimate causal effects. The idea is simple enough to describe without making use of technical jargon.

A/B testing

In the industry it's quite standard to eliminate selection effects by running A/B tests, and the most data-driven companies run thousands of such experiments each year to find causal estimates that drive their decision-making.

I will devote several pages to A/B testing in the Appendix, so I'll just give a very superficial description of the technique here. Our objective is to estimate the causal effect of pulling a lever X on some output metric Y. Say that we wish to quantify the impact that a price discount has on our revenues.

We run an A/B test by splitting our customers into two groups: the A group acts as a control and gets the standard price. In contrast, the B group gets the price discount. Crucially, to avoid selection biases we choose our groups randomly, so that when we compare the average profits across groups, we can rest assured that we in fact estimated the causal effect. I left out all of the interesting technical details, so if you're interested, please consult the Appendix.

Uncertainty

We have now talked about each of the stages in the decomposition: starting with the business, we reverse engineer the actions or levers that impact our objective and corresponding KPIs, mediated by some consequences. However, since decisions are made under uncertainty, this mapping from actions to consequences is not known to us at the time of the decision. But by now we already know that uncertainty is not our enemy and that we can embrace it thanks to the advances in predictive power of AI.

But why do we have uncertainty? Let us first discuss what this uncertainty is not, and then we can talk about what it is. Think about flipping a coin. We know that with a balanced coin the chances it falls on heads are 50% and that the final outcome cannot be fully anticipated from the outset. Since we have played heads and tails for most of our lifetimes, this is an example of randomness that is quite close and natural to us.

This is not, however, the type of uncertainty we have when we are making decisions, and that is good news for us. The fact that ours is not pure randomness allows us to use powerful predictive algorithms, combined with our knowledge of the problem, to select input variables—also known as features—to create a prediction. With pure randomness, the best thing we can do is learn or model the distribution of outcomes and derive some theoretical properties that allow us to make smart choices or predictions.[5]

The four main sources of uncertainty *when we make decisions* are our need to simplify, heterogeneity, complex and strategic behavior arising from social interactions, and pure ignorance about the phenomenon, each of which will be described in turn. Note that as analytical thinkers, we should always know where uncertainty comes from, but it is not uncommon that we end up being taken by surprise.

Uncertainty from Simplification

One of my favorite quotes—commonly ascribed to Albert Einstein—is that "everything should be made as simple as possible. But not simpler." In the same vein, statistician George Box famously said that "all models are wrong, but some are useful." Models are simplifications, metaphors that help us understand the workings of the highly complex world we live in.

I cannot emphasize enough the importance that learning to simplify has for the modern analytical thinker. We will have enough time in Chapter 5 to exercise our analytical muscle through some well-known techniques, but we should now discuss the toll that simplification has.

As analytical thinkers and decision-makers we constantly face the trade-off between getting a good-enough answer or devoting more time to develop a more realistic picture of the problem at hand. We must decide how much uncertainty we're comfortable with and how much we are willing to accept in order to get a timely solution. But this calibration takes practice, as Einstein succinctly puts it in the first quote.

One clear example of the powers and dangers of simplification is maps. Figure 2-9 shows a section of the official Transit for London (TfL) tube map on the left and a more realistic version on the right, also by the transportation authority (*https:// oreil.ly/HONtI*). With the objective of making our transportation decisions fast and easy, a map trades-off realism for ease-of-use. As users of the map, we now face uncertainty about the geography, distances, angles, and even the existence of possible relevant venues such as parks or museums. But to a first approximation we feel comfortable with this choice of granularity since our first objective is being able to get from our origin to a destination. We can take care of the remaining parts of the problem later.

---

5 In the coin tossing example, for instance, after observing the outcomes we may end up modeling the distribution as Bernoulli trials and predict a theoretically derived expected value (number of trials times the estimated probability of heads, say).

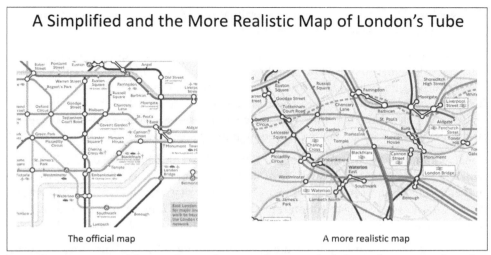

A Simplified and the More Realistic Map of London's Tube

The official map   A more realistic map

Figure 2-9. Sections of the London underground maps—the left panel corresponds to the official tube map, while the right panel shows a more realistic version of the same section

This last point takes me to another related issue: one common simplification technique is to divide a complex problem into simpler subproblems that can each be tackled independently; computer scientists call this the *divide and conquer* technique. When each of these subproblems gives rise to some uncertainty, nothing guarantees that the resulting uncertainty after aggregation becomes more tractable (unless we impose some simplifying assumptions to start with).

The moral of this story is that we should always remember that simplifying a problem usually brings additional uncertainty to the table. As Box, the statistician, commented, "…the approximate nature of the model must always be borne in mind" (*https://oreil.ly/f8ZLH*).

Uncertainty from Heterogeneity

One important source of uncertainty when making business decisions comes from the fact that our customers react in very different ways. This wide variety of behaviors, tastes, and responses can be modeled with the use of distributions since that's how we generally deal with uncertainty (recall Figure 2-5). By doing so we can dispense with the nitty-gritty details of how and why outcomes are so diverse, and just focus on how uncertainty affects our final outcomes. This modeling approach is quite handy and forces us to know some basic properties about distributions.

Take the case of the *uniform distribution*. While it is most commonly assumed for simplification purposes, it can also be used if there's no reason to believe that outcomes will tend to accumulate. To give a concrete example, think about how people waiting for a train during peak hours end up being distributed across the platform. If their goal is to find a seat and enter the train as quickly as possible, it is most natural that they end up distributing uniformly.

We have already encountered the *normal distribution*, which is quite pervasive in the sciences. It is sometimes used for simplification purposes as it has some highly desirable properties (linearity, additivity), but it also arises naturally in many settings. For instance, we may appeal to a version of the central limit theorem (*https://oreil.ly/QvhU4*), which states that under certain conditions, the distribution of averages or sums of numbers ends up being close enough to a normal.

Other commonly used distributions are power-law (or heavy tailed) distributions, which, contrary to the Gaussian distribution, have longer tails.[6] For instance, when modeling the reach or just the number of followers that your influencer has, we may resort to a power-law distribution, but there are many other examples where these distributions arise most naturally.[7]

Figure 2-10 shows the results of drawing one million observations from uniform, normal, and power-law distributions.

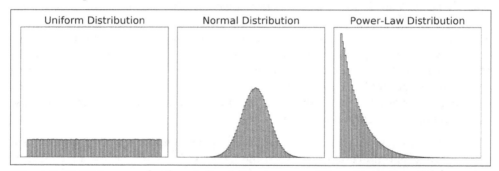

Figure 2-10. Histograms for the results of drawing one million observations from a uniform (left), normal (center), and power-law (right) distribution

Uncertainty from Social Interactions

Another source of uncertainty arises from the simple fact that we are social animals continuously interacting with each other. While this has been taking place for hundreds of thousands of years, the explosion of interactions with modern social networks has made it even more salient and prevalent.

---

6 The normal distribution accumulates 99% of the possible outcomes within 2.57 standard deviations from the mean and 99.9% within almost 3.3 standard deviations.

7 Other examples and applications of power-law distributions in business can be found in Crawford, Christopher G. et al., "Power law distributions in entrepreneurship: Implications for theory and research" *Journal of Business Venturing* 30, no. 5 (September 2015): 696-713. *https://oreil.ly/pSxTh*.

A first source of uncertainty comes from the strategic nature of our interactions with our customers and workforce, just to give two examples. With customer retention offers, for instance, it is not uncommon that customers understand our workings and motivations and end up gaming our system. Similarly, compensation schemes are quite commonly gamed by our sales executives, giving rise to somewhat unexpected results like delayed sales when goals have been or are unlikely to be reached.

But uncertainty may also arise from nonstrategic and very simple decision rules. One well-studied example is John Conway's Game of Life, which evolves in a two-dimensional grid such as the one depicted in Figure 2-11.[8] At any given time, each colored pixel can only interact with its immediate neighbors, thereby creating three possible outcomes: it lives, dies, or multiplies. There are only three simple rules of interaction, and depending on the initial conditions, you can get completely different outcomes that appear to be random to any observer.

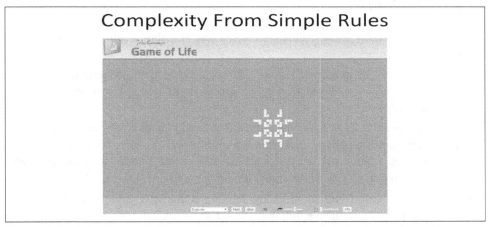

Figure 2-11. John Conway's Game of Life: a plethora of aggregate phenomena arises from three simple rules of how each cell or pixel interact with its neighbors

You may wonder if this is something worth your time and attention, or if it's just an intellectual curiosity. For a start, it should serve as a cautionary tale that even simple rules of behavior can create complex outcomes, so we don't really need sophisticated consumers trying to game our systems. But social scientists have also been using these tools to make sense of human behavior so, at the minimum, they ought to be useful for us when making decisions in our businesses.

---

8 You can "play" the game yourself at *https://playgameoflife.com* and marvel at the rich diversity of outcomes that can be generated by simple deterministic rules. See also *https://oreil.ly/6ruzw*.

Uncertainty from Ignorance

The last source of uncertainty is pure ignorance, as many times we simply don't know what will happen when a lever is pulled, and we are also unaware of the likely distribution of outcomes. In this case, it is not uncommon to start by assuming that outcomes follow a uniform or a normal distribution, later improving our knowledge by some sort of experimentation.

A company's ability to scale testing at the organizational level can create a rich knowledge base to innovate and create value in the medium-to-long term. But there is always a trade-off: we may need to sacrifice short-term profits for medium-term value and market leadership. That's why we need a new brand of analytical decision-makers in our organizations.

Key Takeaways

- *Analytical thinking* is the ability to identify and translate business questions into prescriptive solutions.

- *Value is created by making decisions*: we create value for our companies by making better decisions. Prediction is only one input necessary in our decision-making process.

- *Stages in the analysis of decisions*: there are generally three stages when we analyze a decision: we first gather, understand, and interpret the facts (descriptive stage). We then may wish to predict the outcomes of interest. Finally, we choose the levers to pull to make the best possible outcome (prescriptive stage).

- *Prescriptive decision-making*: decision-making is the act of choosing among competing actions to attain specific objectives. *Data-driven* decision-making is acting upon evidence to assess alternative courses of action. *Prescriptive* decision-making is the science of choosing the action that produces the best results for us.

- *Anatomy of a decision*: we choose an action that may have one or several consequences that impact our business outcomes. Since generally we don't know which consequence will result, this choice is made under conditions of uncertainty. The link between actions and consequences is mediated by causality.

- *Start with the business*: since our aim is to find the best course of action, we'd better be optimizing for the right question. So start with the business. One side benefit is that we usually enlarge the menu of levers available to us.

- *As important as asking the right question is the selection of the metrics to measure the impact of our decision-making*: many data science projects fail not because of the logic used but because we used the wrong set of metrics to measure the impact for our business question. Good metrics should be relevant and measurable.

- *One important skill for us to develop is the ability to create counterfactuals*: since causation mediates the mapping from actions to consequences, we must strengthen our ability to imagine alternative theories of why our business objectives follow from our actions.

- *Estimating causal effects has several important difficulties*: selection biases abound, so directly estimating the causal effect of a lever is generally not possible. We also need to master the use of counterfactual thinking and dealing with heterogenous effects.

Further Reading

Almost every book on data science or big data describes the distinction between descriptive, predictive, and prescriptive analysis. You may check Thomas Davenport's now classic *Competing on Analytics* or any of its sequels (Harvard Business Press), or Bill Schmarzo's *Big Data: Understanding How Data Powers Big Business*, or any of its prequels and sequels (Wiley).

The anatomy of decisions used here follows that literature and is quite standard. We will come back to this topic in Chapter 6, where I will provide sufficient references.

My favorite treatments of causality can be found in the books by Joshua Angrist and Jörn-Steffen Pischke, *Mostly Harmless Econometrics* (Princeton University Press) and their most recent, *Mastering 'Metrics': The Path from Cause to Effect* (Princeton University Press). If you are interested, you can find there the mathematical derivation of the equality between difference in observed outcomes and causal effects plus selection bias. They also present alternative methods to identify causality from *observational data*, that is, from data that was not obtained through a well-designed test.

A substantially different approach to causal reasoning can be found in Judea Pearl and Dana Mackenzie's *The Book of Why: The New Science of Cause and Effect* (Basic Books). Scott Cunningham's *Causal Inference: The Mixtape* provides a great bridge between the two approaches, focusing mostly on the first literature (econometrics of causal inference) but devoting a chapter and several passages to Pearl's approach using causal graphs and diagrams. At the time of writing, this book is free to download on his website (*https://oreil.ly/4FfDp*).

I will provide many references on A/B testing in the Appendix. My discussion of uncertainty follows many ideas in Scott E. Page's *The Model Thinker: What You Need to Know to Make Data Work for You* (Basic Books). This is a great place to start thinking about simplification and modeling, and provides many examples of when and where distinct distributions, complex behavior, and network effects appear in real life.

## CHAPTER 3

Learning to Ask Good Business Questions

Chapter 2 provided a quick overview of the general framework we'll be developing in the upcoming chapters. Since our ultimate objective is to translate business problems into prescriptive solutions, we should start learning how to ask the *right* questions. I hope it doesn't come as a surprise that learning to frame the questions can have an impact comparable in magnitude to adopting the techniques that will follow.

We also introduced a very simple technique that I've found quite useful to understand what we really want to accomplish: *the sequence of* why *questions*. You start by questioning what you think you are trying to accomplish, then move up one level or stop when you are convinced that the business objective is in fact just right. In our voyage to find prescriptive solutions, it is of the utmost importance to guarantee that we are tackling the right objectives. One nice byproduct that will be quite handy in Chapter 4 is that this usually enlarges the set of possible actions or levers we have. This is usually the case when we start by questioning an action and the procedure ends up taking us to the metrics we really want to affect. It is almost natural, then, to question if there are other actions that can be used to impact the same objective.

In this chapter, we will delve a bit more into some of the best practices when asking good business questions; understand the difference between descriptive, predictive, and prescriptive questions; and we'll end by providing some examples from common use cases. These were selected from my own experience, from other use cases I have discussed with both students in class and colleagues, and because they are good to present and understand the methods. But first, we should better understand where business questions come from (Figure 3-1).

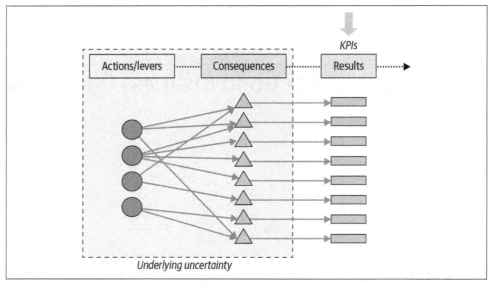

Figure 3-1. Start with the business

From Business Objectives to Business Questions

Most companies are organized vertically by clearly separating each area's responsibilities and objectives (Figure 3-2). In the past few years, however, the *Agile* movement has helped many companies to break the functional silos and organize into cross-functional teams. The outcome has been that each team has very clearly delimited business objectives and metrics to pursue.[1]

---

1 On different organizational structures see *https://oreil.ly/T1MzW* or *https://oreil.ly/EOa_D*.

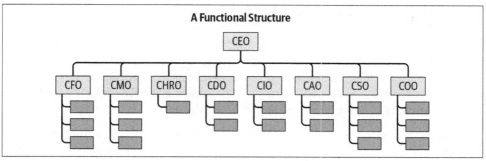

Figure 3-2. An example of a company organized by functional divisions—from left to right, the acronyms correspond to Chief Officers in Finance, Marketing, Human Resources, Data, Information, Analytics, Sales, and Operations, respectively—and there are many more such acronyms

This is good news for us, since our top-down business objectives are usually well-defined and, supposedly, relatively easy to evaluate with the use of appropriate KPIs. It is our task, however, to ask the necessary business questions to achieve these objectives. In general, for any business objective there are multiple business questions that can be asked, and for each of these there are different actions or levers.

Hard and Soft KPIs

Even though there isn't an accepted definition, it not uncommon to hear about *hard* and *soft* KPIs. Hard metrics are thought of as being relatively straightforward to measure objectively, like financial KPIs, for instance. On the other hand, soft metrics like brand awareness, customer satisfaction, or service quality are more difficult to measure in an accurate, objective manner.

The distinction isn't obvious, and there will always be ground for debate, but there is a sense that financial metrics rest on firmer ground and are more precisely measurable. Recall from our discussion in Chapter 2 that measurability is one of the properties we look for in good KPIs.

How do we formulate good business questions? Since for our purposes a business question is always *actionable*, it is first necessary to understand the business objectives we want to effect as well as the metrics used to assess the results, and to at least have some idea of some candidate levers we can pull. If you have not identified any actions you can take, either the question is not actionable, or you haven't thought through the problem. Otherwise we are on the right track. We now need to distinguish between descriptive, predictive, and prescriptive questions.

Descriptive, Predictive, and Prescriptive Questions

In their article "What is the question?" (*https://oreil.ly/xcPM7*) Jeff Leek and Roger Peng describe six types of questions that you may want to answer with data: descriptive, exploratory, inferential, predictive, causal, and mechanistic. Data analysis usually mirrors our analytical processes, so these map somewhat neatly to the threefold classification used here: descriptive, predictive, and prescriptive.

In Chapter 2, I described the three types of analysis, so here I'll just repeat that *descriptive* analysis generally looks at the past, *predictive* at the future, and *prescriptive* finds the best actions we can take today to change the future.

One of the motivations to write this book was the casual finding that most people tend to ask descriptive questions and have trouble finding the right place to use predictive and prescriptive analysis. Later in this chapter, I'll provide enough examples to eliminate any confusion you may still have about these concepts.

Always Start with the Business Question and Work Backward

One of the preferred catchphrases in the data world is that practitioners create value by finding *actionable insights*. While there's nothing wrong about this assertion, there is a risk of spending hours, days, or weeks in search of the million-dollar insight.

At some point in my career, I did something similar: I found that it is relatively easy to look at the tails of the different distributions—those with a lower probability of arising—and find unseen business opportunities in these microsegments. Since most models focus on the average customer (thereby neglecting the tails), this was a relatively straightforward way to help my employer make some money. That was the definition of low-hanging fruit. There were two problems, however: it was not scalable, and it was a highly manual and time-expensive process.

In general, a better practice is to always start with the business question and move backward to the data. This process leads to faster actionable insights, since you have already started with the actionable insights you want to find from the beginning![2] The process described in this book will help you tighten and refine the analysis, and hopefully you won't waste your own or your team's valuable time in the search for the promised actionable insights.

---

2 Cassie Kozyrkov, Chief Decision Scientist at Google, has presented a similar view in "Never Start with a Hypothesis" (*https://oreil.ly/iMeVy*) and "The First Thing Great Decision Makers Do" (*https://oreil.ly/uw7bO*).

Further Deconstructing the Business Questions

The sequence of *why* questions helps us move from specific to more general questions in the quest to find the metric that we really want to impact. The risk is that this final metric may be too general to be actionable (the highest level is almost always something like "increase profits"). We should remember, however, that our own business objectives act as a natural constraint, so there's usually an upper bound in the sequence. Furthermore, there are techniques that allow us to do just the opposite and start decomposing questions in order to find just the right level where we can clearly identify intermediate objectives that are also actionable.

For instance, consider the problem of finding the best actions to get the highest conversion rate possible for your outbound marketing campaigns. Notice that I have already framed the question as a prescriptive one on purpose: the business metric is well defined (conversion rate), and if we find suitable actions then we can (in principle) choose the best ones for our purposes.

Decomposing Conversion Rates

Any ratio can be decomposed by multiplying and dividing by different metrics. Here we start with the ratio of sales to leads—the conversion rate—and first multiply and divide by the number of reached customers. We then repeat with the number of customers that we actually called (dialed). In the end we reorganize the equation so that each of the parts represents a relevant metric in its own right.

$$\underbrace{\frac{Sales}{Leads}}_{CR} = \underbrace{\frac{Sales}{Reached}}_{A} \times \underbrace{\frac{Reached}{Dialed}}_{B} \times \underbrace{\frac{Dialed}{Leads}}_{C}$$

Conversion rates have the property that they can be easily decomposed, leading to more directly actionable questions as we'll now see. In this case, the conversion rate (CR) is the product of three different ratios, each with different possible levers to pull, and with possibly different accountabilities.[3] Starting from the rightmost ratio (C), if out of 100 leads you only tried to contact 15 by dialing their numbers, it could mean that your telemarketing team is in a low productivity valley, and you'd better talk with their manager to find actions or at least understand what's happening.

Similarly, if you have already dialed each of the phone numbers and were only able to reach a low fraction of them (B), you may want to search for variables that allow you

---

3 In case you didn't notice, I'm multiplying and dividing by the same metric so that the equality is always preserved.

to predict the best time to contact your customers: this may now be a job you assign to your company's data scientists.

Finally, if your sales team is only able to convert a small fraction of those who were reached (A), it could be that the predictive models should be improved to generate higher quality leads, that your compensation scheme needs to be adjusted, or that your product-market fit is not right yet.

Notice how the decomposition immediately allows us to find intermediate metrics or questions, with their corresponding actions, to increase conversion rates. This trick can be easily applied to most conversion funnels. Let's take the example of an archetypical *two-sided platform*.

Example with a Two-Sided Platform

Two-sided platforms, or marketplaces, generally try to match users on one side with users on the other side. Facebook, for instance, matches companies that want to place ads (in order to make sales) with the right customers (users of the social network). Amazon tries to match distributors or sellers of goods with the right buyers, Uber matches drivers with passengers, and so on.

Imagine you start your own dating platform. Here the two sides are users that want to find their perfect match. Most of these dating apps allow users to communicate with each other. For simplicity let's say that the rules of the game allow only one message per user; the more general case will only make the decomposition longer.

After exchanging messages, if users like each other, they can take it to another place (a coffee shop, a bar, or a restaurant). Your team of data scientists wants to improve the app's matching efficiency, measured by the ratio of converted matches. For the sake of the argument, let's say that users always provide feedback to the app so that we can always know whether two users met.[4]

We have a dataset of all users, their interactions (Message 1 and Message 2), and the final outcome (Met or Didn't Meet). The matching efficiency (ME) can then be decomposed as follows:

---

4 Some dating apps actually incentivize users to provide this formal feedback, but other times there are indirect ways to measure the matching efficiency.

Decomposing the Matching Efficiency for a Dating App

We display images of users in a dating app with the hope that these are high-quality potential matches for other users. Each user may decide to start a conversation by sending a first message (message 1) which may be replied to by the second user (message 2). After this they either decide to meet elsewhere or stop the conversation.

$$\underbrace{\frac{\text{Met}}{\text{Displayed}}}_{ME} = \underbrace{\frac{\text{Met}}{\text{Message2}}}_{A} \times \underbrace{\frac{\text{Message2}}{\text{Message1}}}_{B} \times \underbrace{\frac{\text{Message1}}{\text{Displayed}}}_{C}$$

In this equation, each term denotes the number of occurrences for each event. For instance, *Met* denotes the number of people that ended up meeting, and *Message1* and *Message2* denote the numbers of first messages sent and the number of replies, respectively. Each ratio should be less than one, since the number in the numerator is a count for a subset of the event in the denominator. This is always the case when decomposing conversion funnels.

Notice what the decomposition tells us: if we want users to match, we need them to exchange messages, which can be represented by three ratios: once a user finds someone displayed on the app, they can send a first message (Message1). This ratio (C) shows if the algorithm is being efficient from the point of view of user 1: if the app displayed 10 candidates and all were of high quality, then they would message all of them.[5] User 2 may now reply or not: if they do, it may signal that the algorithm is doing also a good job for them (B). Finally, after the second message is delivered, the two either meet or do not (A).

But not everything depends on the algorithm's accuracy. A decision to start a conversation (Message1) or reply (Message2) also depends on each user's attention, say, because of delays in communication: dating apps are fast-moving platforms, so if any user takes too long to reply, the other user may lose interest and continue searching for potential dates. We can then devise methods to incentivize faster communication (emails, push notifications, or pop-ups reminding users that someone is waiting for a reply). Bumble, for example, does just that: the initial contact for each side must be within the first 24 hours or the match is lost.

---

5 This is not to say that there aren't different strategies: a user may start only messaging the top candidate displayed and see where that takes them. In that sense, you may want to restrict the analysis to a decomposition that excludes the last ratio.

The takeaway here is that some business questions can be further decomposed to find the right actions, so we may need to reconsider effecting intervening KPIs to achieve our objectives. We will now go through some real-life common use cases.

Learning to Ask Business Questions: Examples from Common Use Cases

We will now go through a selection of examples, starting with what I've seen is the standard way to frame the business question, then posing the corresponding descriptive, predictive, and prescriptive counterparts. Recall that a good prescriptive question should always find ways to pull some levers so that we get the best possible outcome in terms of the business objective we have chosen. I will further develop some of these examples in subsequent chapters, to the point of providing what I think is a good-enough prescriptive solution; I'll let you find ways to improve on that. For now remember that our purpose in this chapter is just to learn to translate business questions.

Lowering Churn

In all companies we need customers in order to generate revenues. We start by *acquiring* customers and then part of our job is to keep them loyal for the longest time possible. The rate at which customers leave—the churn rate—is the ratio of the number of customers we lost in a fixed period of time relative to the overall customer base in that same period. Since acquisition costs can be relatively large compared to retention costs, most companies have specialized areas with the specific objective of safeguarding as much as possible their current customer base.

This is one standard use case in most companies, so it provides a great way to start applying the techniques (Figure 3-3).

Lowering Churn

		Value potential
Business Question I want to lower the churn rate in our company	Why? What KPIs am I trying to impact?	
Descriptive Question How many customers are we losing? Who are they?	Are all customers the same?	$
Predictive Question Can I know in advance which customers will leave?	What characteristics make some customers more likely to churn?	$$
Prescriptive Question If I can anticipate those customers that leave, what type of retention policies should I use and how should I assign them?	What KPI will I impact? Short/long-term profits? What are the costs of each retention action?	$$$$

Figure 3-3. Different questions for the case of customer churn

Defining the business question

Let us start with the business question most companies face: how can we lower the churn rate? This is an example where we start with an action and not with the business objective, so we can apply the sequence of *why* questions, and most likely we'll end up with the simple fact that customers are our main source of recurring revenues. It seems straightforward, but this simple fact takes us to the main KPI we want to maximize: it's not the churn rate that we want to make as small as possible, it is revenues that we want to be high. Or is it? You can always give away everything to keep your customers, thereby increasing your costs. It follows that this is *not* the right metric we want to impact either: it is profits, measured as the difference between revenues and retention costs.

Descriptive questions

At the most descriptive level, we want to do several things. Of course, we start by asking whether our churn rate is abnormally high and how it has evolved in the past. We may start at the most aggregate level by looking at time trends and patterns of seasonality, which gives us a sense of our current health status. But data has the power to go deeper and tell us *who* the customers are that have already left. Are they high- or low-value customers? What is their tenure with us? Have they contacted us in the past to show their dissatisfaction? Are they geographically located in specific areas? What are some of their sociodemographic characteristics, such as age and gender? What are their usage or consumption patterns?

We can get as granular as our data and time allow. But you get the idea: this is just a snapshot, and hopefully I have convinced you by now that no matter how high-definition it is, it's hard to get more value out of it. At this point it has mainly been informative. The real value from this descriptive analysis is its ability to take us further in our quest to find the best decisions we can, in order to achieve our ultimate objective.

Predictive questions

AI and machine learning can help us find answers to the predictive question: can we anticipate which customers are more likely to leave? Thanks to the richness of our descriptive analysis, we have hopefully now found some of the primary drivers that explain our current churn rate. But data alone can only take us so far. The best data scientists are those who *understand* and *hypothesize* why customers are leaving. In this way they can create more specific predictors in a process called *feature engineering*, which is the best way to get really good predictive power. Knowing what to include or not in our models is the holy grail in the construction of good models, even more than, say, choosing the ever-more powerful available algorithms.

How much value does the predictive stage provide? In Figure 3-3, I suggest it's higher than the value from the descriptive step, but it could very well be null or even negative, as discussed in Chapter 2.

Prescriptive questions

Finally, we have arrived at the prescriptive question: what levers should we pull if we want to maximize our profits from our retention campaigns? But are we thinking of *short-term profits*? Will customers learn our strategy and start gaming our retention system, thereby increasing longer-term costs? Most mature companies prefer to use the customer lifetime value (CLV) we introduced in Chapter 2, and I agree that this is indeed a better picture of the long-term net value of our customers. But this choice of metric comes with its own set of difficulties: to paraphrase Yogi Berra, the future is hard to predict, and it is even harder to understand the longer-term effects of our actions.

We will talk about levers in Chapter 4, but suffice it to say here that for the case of customer retention, we can always give away something at least in the form of discounts. What, then, are the right discounts for each customer? The CLV provides an upper bound on how much we should reasonably give away, but we always want to find the action with the lowest cost that guarantees retention. This takes us closer to the personalization of levers.

The prescriptive ideal is one where we choose the *right* action, at the *right* time, for the *right* customer. Too much *right stuff*: prescriptive analysis is hard, so most times we will try to simplify the problem. I will talk about the power of simplification in Chapter 5. But at least we have already framed the question in a way that, by design, can potentially generate the highest achievable value. Recall that this chapter is about learning how to frame questions. In Chapter 7, I will go into the details of one possible solution to this use case.

Cross-Selling: Next-Best Offer

Most companies sell more than one product or offer more than one service. Economists call the natural advantage that a company may have when offering products that can benefit from similar production processes *economies of scope*. It is thus logical for most of us to look for ways to deepen our relationship with our customers by trying to do some cross-selling. In consulting jargon, it has been relabeled as the now-famous *next-best offer*, which already takes us to the prescriptive terrain. That label notwithstanding, we'll now see that it is far from obvious what business objective is being optimized here.

Cross-Selling: Next-Best Offer		Value potential
Business Question What should I offer now to each of my customers?	Why? What KPIs am I trying to impact?	
Descriptive Question What are the patterns of consumption of our customers? Do some of these patterns arise more naturally? With which customers?	Are all customers the same? Are all sequences of products equal?	$
Predictive Question Given each customer's previous purchases, what are they more likely to buy now?	What characteristics make some customers more likely to buy each of our products?	$$
Prescriptive Question If I want to get the highest value from each customer, what is the best offer I can make at each time?	What KPI will I impact? Should I think in terms of individual products or sequences of products?	$$$$

Figure 3-4. Different questions for the case of cross-selling

Defining the business question

The business question here is straightforward (Figure 3-4): *what should I offer now to my customers?* If you wonder why would you even want to do such a thing (the sequence of *why* questions), the answer is not as clear as with customer churn. The difference here is that cross-selling has two effects. The direct effect is the usual channel of higher revenues and profits. But the indirect channel is more interesting and complex: customers who buy more from us tend to be more loyal, thereby increasing the time they remain our customers. Because of this, many times we may consider cross-selling at a discount just because the long-term *overall* profits are higher, even when the transaction of an individual product is made at a loss for the company. It appears, again, that CLV is the right metric to optimize.

Descriptive questions

On the descriptive terrain, the type of questions one would normally explore are things like the patterns of consumption for different customers. Specifically, it is natural to explore if certain *sequences* of products arise more naturally with different customers. Think of a bank, for example: most customers start at a young age with a relatively simple product like a credit card. With time, and with their incomes increasing with job experience, they tend to move to more sophisticated credit and investment opportunities: you may first get a mortgage, move to life insurance, and so on. With sequences, the order in which each product is purchased matters, so it is standard to start by looking for those patterns in the data.

Predictive questions

Since each customer has already purchased something, it seems natural to ask if we can *predict* what they are most likely to purchase given their patterns of consumption to date. We could then move proactively and not wait and see if they purchase with us or our competitors. But should we offer the product with the largest profits for us, even if it's highly unlikely to be purchased? Going back to the bank example, you may want your customers to take out a mortgage loan (because of its large returns), but for college students and young professionals it may be highly unlikely that they will accept such an offer. This takes us to one of the most interesting trade-offs in next-best offer analysis: likelihood of purchase versus increase in value. Which in turn brings us to the prescriptive question.

Prescriptive questions

Since we can offer several items to each customer, which one should we select so that we can capture the highest value? As mentioned previously, since we are dealing with sequences and time, the right metric is most likely the CLV. In a truly customer-centric sense, the prescriptive ideal would take us, again, to find the *right* product, for the *right* customer, at the *right* price, and the *right* time. Later, we'll see an approach to tackling this highly complex question.

CAPEX Optimization

Automotive, oil and gas, telecommunications, and airlines are examples of industries that are capital intensive: in order to operate, they need to allocate large amounts of resources to building and maintaining factories and plants, towers, planes, and any other physical assets that depreciate in time. This type of investment is called capital expenditure, or CAPEX, and is common to all industries, not only the four cited previously.[6]

One natural question that CFOs and other executives have in any company is how to allocate CAPEX, say, across functional areas or geographical locations (Figure 3-5). Since it may represent a large part of a company's cash flow, we even have specific KPIs to measure its impact, such as the Return on Investment (ROI) or Return on Capital Employed (ROCE). Nonetheless, we should always proceed to question why we need to allocate CAPEX and what, exactly, we are trying to accomplish: for instance, where is income in the ROI numerator coming from?[7]

CAPEX Optimization

		Value potential
Business Question How should I allocate capital expenditures this quarter?	Why? What KPIs am I trying to impact?	
Descriptive Question What has happened in the past when I invest more in some geographies?	Can I believe in those correlations?	$
Predictive Question If I invest in geography X will sales or customer satisfaction increase? By how much?	Are these casual effects?	$$
Prescriptive Question If I want to maximize returns on investment, how should I allocate all of my CAPEX across geographies?	Is this really the best I can do? Can it be that this is one of several better allocations? Can we do better?	$$$$

Figure 3-5. Different questions for the case of CAPEX optimization

At a descriptive level, we could start by finding correlations between different CAPEX allocations and revenues across geographies. This exploits the variation in previous investments with the key metric that we believe is impacted. Another possibility is to exploit the variation in time and plot aggregate series in search of any preliminary hints of a relationship between CAPEX allocations and revenues. If we don't find such evidence, we could change our objective and think about variation in profits, as CAPEX may also be aimed at reducing costs.

The main problem we have when considering any investment is that we *do not know* what the returns will be, so it would be great if we could perfectly predict them.

---

6 Compare this with operating expenditure, or OPEX, that includes, among many other things, the salaries paid to your employees.

7 Recall that $ROI = \dfrac{\text{Income from Investment - Cost of Investment}}{\text{Cost of Investment}}$

Optimal allocation could then just be a matter of rankings: if I have one dollar to invest and know the returns of all candidate allocations, I would put it on the one with the highest returns. But can we trust the correlations in our descriptive analysis? Is the effect we find really causal? As usual, the hard part is to find reliable causal predictions, and that's what our data scientists will try to find with the use of their machine learning toolkit.

But assuming we have achieved reliable and accurate predictions, the prescriptive part is almost done for us: allocate your budget in different geographies ranked by their returns. Later I will show you one example of how this can be done, but for now all we need to learn is the framing of the question.

Store Locations

One of my favorite use cases is where to open a store, and since we have already talked about CAPEX optimization, we immediately see that this is just an instance of the same problem. We have a budget to strengthen our commercial presence and ideally we would just open a store where we will have the largest possible return (Figure 3-6). A natural KPI is the net present value (NPV) of the store's profits, or is it?

Just to demonstrate the complexity of the problem, consider opening a store that is already very close to another one (have you ever wondered why there are so many Starbucks in one specific block or neighborhood?). You could capture extra revenues and profits but only at the expense of profits in nearby stores. So a more reliable KPI would be the aggregate level of profits, at least at a local (neighborhood, street, or even city) level.

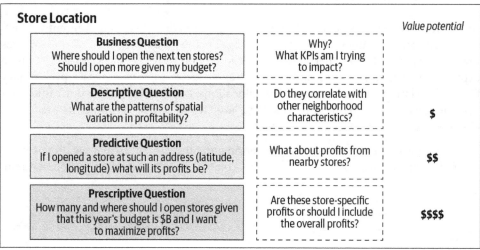

Figure 3-6. Different questions for the case of where to open a new store

Descriptively, I would start by looking for patterns of variations in profits across different spatial locations: are there any of our own stores in the vicinity? What about the competition? Do we have data to approximate the number of potential customers that enter different stores? What about the average income in the neighborhood? Is it a residential neighborhood? We may pose many questions in order to find the patterns that explain variations in profits.

Just like with capital expenditure allocation, if we could perfectly predict the NPV of overall profits, we would be almost done: allocate all of your budget to opening stores ranked by this KPI. Of course, I'm assuming here that you have a finite budget and that you won't invest in opening stores that have negative returns.

Who Should I Hire?

It is an understatement that our employees make our company great or not so great. So one of the most important decisions we constantly make is who to hire. Human resources units spend considerable efforts in having a robust and reliable recruitment process (Figure 3-7). The main problem we face in hiring is that some of the KPIs may not be as easy to measure. Consider productivity, for example. If you are a salesperson we can clearly measure your own productivity with the number of sales in a fixed period. But for many other positions it is harder to measure productivity or even an employee's impact on revenues.

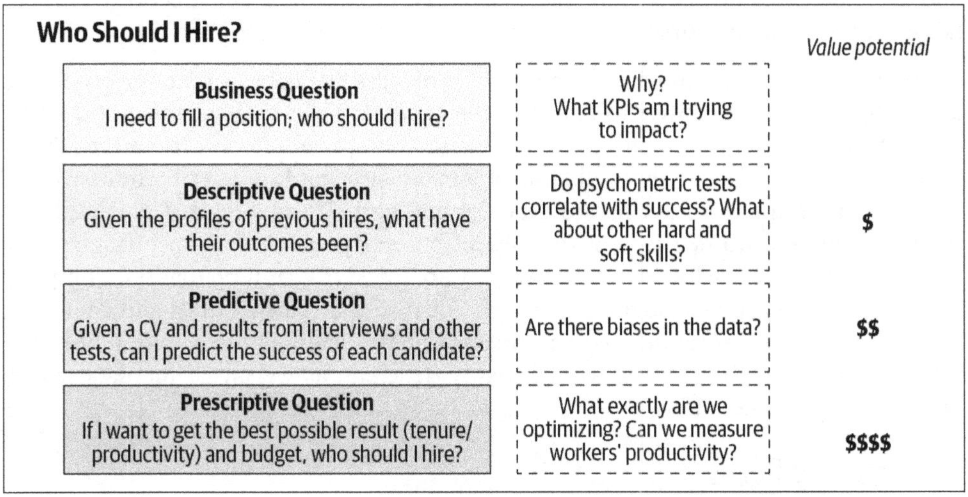

Figure 3-7. Different questions for the case of hiring decisions

Suppose we can reliably measure productivity like in the case of our sales force. Is that the only KPI that matters? What about tenure? You may not want to hire a superstar salesperson if she changes jobs a month later, as her sales may not compensate for the hiring and training costs. Ideally we would like to measure something like the customer lifetime value, so let's use the analogous term—the employee lifetime value: the net present value of our individual contribution to profits. That way we can include the expected duration and the financial impact.

Let us imagine we have a dataset for all of our salespeople in the past 24 months. As with customer churn, we need our dataset to include active employees with different tenures as well as those who have already left. This variation would allow us to start finding patterns. It would also be ideal to have some of their hard performance metrics (monthly sales) and also softer metrics like their 360 survey results. Finally, it would be great to have some of the data we actually get when hiring: their CVs, education, previous experience, psychometric studies, gender, age, etc. We can then search for correlations in the data to have a sense of the kind of variables that may predict performance.

The predictive question is simple enough: with that ex-ante information—anything we collect before making the hiring decision—can we predict the candidates' performance? If we could, the problem would be at least solvable, but as before, there are complexities. The main issue here is one related to all search problems (think of the problem of finding a significant other): should we keep looking for new candidates? Is this the best we can find? If we keep searching for another month, say, will we be able to find someone better? We will talk later about the *explore-exploit* trade-off common to most search problems, but for now just notice that we can either hire and get the best of our new employee—that's the exploit part, but it is not related to labor exploitation, of course—or we keep exploring the market for better candidates.

One final word of caution when using AI in problems like these is that our predictive algorithms are very sensitive to biases in our data, and our data scientists should go through the trouble of searching for them and trying to find ways to correct them. Imagine that your dataset shows that most female salespeople are very productive but quit after a month. Your prediction model might end up showing that the employee lifetime value for women is considerably lower, and you will end up hiring mostly male candidates. But why are women leaving so quickly? Can it be that our company has a terribly misogynist manager? We should fire that person first, and then hire more women. The moral here is that unless we have debiased our data the best that we can, our predictive models will be highly deficient: as we say in the data world, "garbage in, garbage out."

Delinquency Rates

Many companies provide ways to finance their customers' purchases, for instance, with store-specific credit cards that are most commonly found at large retailers. Even better if we can leave that job to specialized firms (banks), but many times we do the funding ourselves. The business question is how to provide lending without increasing the delinquency rates (Figure 3-8).

Delinquency Rates

Value potential

Business Question	Why?	
How can I increase loans without increasing delinquency rates?	What KPIs am I trying to impact?	
Descriptive Question	Can I separate individual from aggregate effects?	$
What patterns in delinquency rates have we had in the past?		
Predictive Question	Are there biases in the data? What about adverse selection?	$$
For any given customer, can I predict the likelihood of delinquency?		
Prescriptive Question	If a customer defaults, how much can I recover?	$$$$
To whom, how much, and at what interest rate should I lend money?		

Figure 3-8. Different questions for the case of lending decisions

The descriptive questions are similar to the cases we have already discussed, but let me just reinforce two ideas: if you have correctly defined the business question and framed it as a prescriptive one, you should be looking for patterns in the data that guide that objective, and not the other way around. Also, we should try to exploit variations in the underlying characteristics of the problem in order to predict the outcome we care about. So I would look for variation across geographies and customers and correlate them with delinquency rates and delinquency outcomes to set up the predictive problem: can we predict if a customer will default on a loan? If they do, can we recover any fraction of it, possibly through an aggressive collection strategy?

We will talk more about ethical problems later, but again, it is important to mention that biases in our data can pervasively affect the outcomes we want to pursue. With loans we should be careful not to hurt underrepresented groups because of the way we have lent in the past.

To set up the prescriptive question, start with the metric we want to affect: it's not whether a customer defaults (their default probability), but rather the expected benefits net of costs from the loan. We will deal a lot with expected values later, but for now it suffices to say that to formulate and provide answers to all prescriptive questions, we need to have a clear understanding of the costs and benefits from our actions. What are our levers? We are certainly able to determine the amount of the loan, and of course, the decision to lend or not. Banks also have the ability to set up and change the interest rate, but because of regulatory issues this lever may not be available to other companies.

Stock or Inventory Optimization

A very common problem for most companies is how many units of each product should be in each store's inventory. In a similar vein, banks constantly decide how much cash to have in their ATMs. Let us start backward this time, starting with the prescriptive question (Figure 3-9).

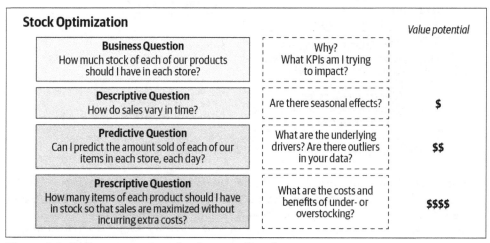

Figure 3-9. *Different questions for the case of stock optimization*

Consider the costs and benefits from over- or understocking one particular item. On one level, if there are not enough units we will hinder our sales in a given day, so the cost is reduced revenues. Understocking can also increase transportation and logistics costs that may be considerable enough to include in our analysis. What about over-stocking? There is some probability the value of those items may decrease, either by depreciation, by the risk of mismanagement or theft, or just because tomorrow a new and better alternative arises and there will be no demand for the old stuff.

At the prescriptive stage, we would want to find the right amount of each item to minimize the *expected* sum of these costs. Later I will delve into the details of how this can be done. What is the underlying uncertainty of the problem? We will use AI to help us deal with it in the predictive stage.

First of all, we do not know how many units of each product will be sold. If each day we always sell the same amount, say 100 units, at the very least we should always have those 100 units. A dissatisfied customer will not only represent the current foregone sales opportunity but possibly many in the future: they and their acquaintances may not come back to the store, so let's hope they're not an influencer! Therefore, we should start by predicting demand in a fixed period of time. But what is that period? A day? A week? It depends on all other costs: if transportation is cheap relative to other risks such as theft or depreciation, you can stock again tomorrow without a problem (that is the case with ATMs, for example). Otherwise you may need to predict the likelihood of these events taking place. Again, we are in the arena of expected values, a topic to be discussed in a later chapter, but as you can see, optimization can be very hard.

So now we can guide the descriptive analysis: how do sales vary across time and geographical locations? Are there seasonal effects? What about theft and robberies? If your items are durable goods (cars, fridges, cell phones, laptops, or the like), how do their values fall with time? You can see the picture now.

Store Staffing

Our final example has to do with the problem of choosing the number of salespeople in a store (Figure 3-10). In a sense it is a similar problem to the stock problem: what are the costs and benefits of over- or understaffing?

Salespeople per Store   *Value potential*

Business Question	Why?	
Can I make smarter staffing decisions in each store?	What KPIs am I trying to impact?	
Descriptive Question	And costs?	$
How do customer satisfaction and sales correlate with our staffing choices in the past?	Do they vary by store?	
Predictive Question	How do we measure	$$
If I have X salespeople in store Y, what will customer satisfaction, sales, and cost be?	customer satisfaction?	
Prescriptive Question	Are we oversimplifying? After making our new	$$$$
How many salespeople should I hire each day to increase profits and customer satisfaction?	decisions will we be able to assess the results? What is operationally feasible?	

Figure 3-10. Different questions for the case of staffing decisions

If we do not have enough people, we will certainly have lower revenues: those customers who wait a long time will leave the store and buy from the competition. Or they will stay and buy today, but their lower satisfaction will likely result in higher customer churn, affecting our revenues in the future. On the other hand, overstaffing creates unnecessary and foreseeable costs. Therefore, expected profits—revenues from sales minus the staffing cost—seem to be a reasonable KPI to optimize. The customer churn effect may be just too much to begin with, so let us start by trying to find the right number of salespeople in our stores to have the highest possible expected profits in a day. If we tackle this already difficult problem, we can proceed to optimize the longer-term version (recall the value of simplification).

What do we need to know in order to solve this problem? How would we proceed if there is no underlying uncertainty? Ideally I would like to know the number of customers coming to our store at any given time in a day, say, each hour or in time periods of 30 minutes. We will need to predict this flow of customers in each of our stores. This will naturally lead us to waiting times given the size of our sales staff. We may now need to decide what is a reasonable waiting time as the limit of no waiting time may just be too costly, especially since there are peak hours where we have many customers and valleys where it appears that we have overstaffed. Waiting times affect our profits today.

What should we look for in the data, then? Variation across stores in demand and staffing is what we need to exploit, but we should also consider the outcomes we want to predict: sales, profits, waiting times, and customer satisfaction are four that immediately pop up. The descriptive stage should be set up to search for these correlations.

Key Takeaways

- *Business objectives are usually already defined*: but we must learn to ask the right business questions to achieve these objectives.

- *Always start with the business objective and move backward*: for any decision you're planning or have already made, think about the business objective you want to achieve. You can then move backward to figure out the set of possible levers and how these create consequences that affect the business.

- *The sequence of* why *questions can help define the right business objective you want to achieve*: this bottom-up approach generally helps with identifying business objectives and enlarging the set of actions we can take. But other times you can also use a top-down approach similar to the decomposition of conversion rates.

- *Descriptive, predictive, and prescriptive questions*: descriptive questions relate to the current state of the business objective; predictive questions are about its future state. Prescriptive questions help us choose the right levers to attain the best possible future scenario.

Further Reading

I haven't found any books on how to ask *good* business questions in the context of decision-making. This is not to say that I'm the first practitioner that suggests that the way we frame our business problems can make a big difference in the results. Almost any book on data science methodology will at least mention the topic. You can go back to the references in Chapter 1 or check Foster Provost and Tom Fawcett's *Data Science for Business* (O'Reilly).

In my opinion, the literature has at least two shortcomings: most data scientists rarely care about solving the prescriptive problem and would rather focus on providing high-quality predictive solutions. Also, the literature directed to business people hasn't been able to provide end-to-end views of decision problems that can be tackled with AI and analytical thinking.

Several of the use cases covered here can be found elsewhere, but I'm not sure at what level of detail. You can search online for white papers written by consulting firms and they will provide some possibly interesting insights (but of course, consulting firms make money from developing those use cases, so don't expect too many details).

On two-sided platforms, I enjoyed reading Geoffrey G. Parker, Marshall W. Van Alstyne, and Sangeet Paul Choudary's *Platform Revolution: How Networked Markets Are Transforming the Economy and How to Make Them Work for You* (W. W. Norton & Company) and David S. Evans and Richard Schmalensee's *Matchmakers: The New Economics of Multisided Platforms* (Harvard Business Press).

The topic of ethical concerns in machine learning is an important one, and I will provide references in Chapters 6 and 8.

## CHAPTER 4

Actions, Levers, and Decisions

Chapter 3 was all about learning to translate business problems into prescriptive questions that, in our case, must always be actionable. But what is actionable? Or even better, is *everything* actionable? We now turn to this question in our quest to find levers that take us closer to the prescriptive ideal.

One word of caution: to find levers, we need *to know our business*. This is not to say that you must have spent many years in one specific industry. That might help, as you must've developed strong intuitions about why things work and when they don't. But it is also true that many times having a non-expert, even naive, view can help us think out of the box and expand our menu of options.

Going back to our decomposition, we will now move from the outer right side, where business outcomes live, to the outer left side, which contains the levers we pull (Figure 4-1). As we've already mentioned, this is the natural and healthy sequence to adopt: we start with the business, and then ask how we can achieve the best results by pulling the right set of levers.

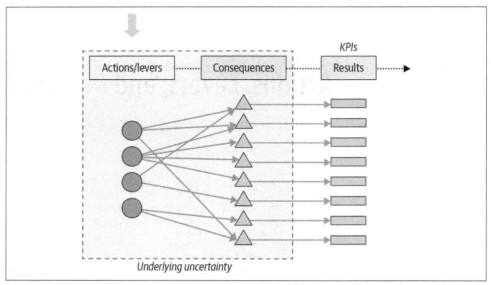

Figure 4-1. Identifying the levers we want to pull

Understanding What Is Actionable

The hard truth about life and business-making is that most of our objectives can only be achieved indirectly, through actions we take. For instance, we can't increase our sales, or productivity, or customer satisfaction, or reduce our costs just because we say so. Intervening factors (human or technological) restrict our ability to do all that we want to accomplish.

The impact that our decisions have on our business objectives is mediated by the rules of cause and effect, and it usually takes a lot of experimentation and domain knowledge to understand what works and what doesn't for our businesses.

In general, we can divide levers into two types: those that depend mostly on the rules of the physical world to create consequences and those that arise from human behavior. As you would expect, each of them has their own sets of complexities and difficulties. Levers of the physical type depend on our understanding of the laws of nature and on technological advances. Human levers depend on our understanding of human behavior.

Physical Levers

As it turns out, the original use of the word "lever" is of a physical nature: you take a beam and a fulcrum, pull the beam down, and you are now able to move objects that were too heavy to lift by yourself. This use notwithstanding, physical levers have become a landmark of modern economies: the rapid growth during the Industrial Revolution, the invention of the microchip, and the current internet revolution, just to mention a few, were vastly facilitated by this class of levers.

Thanks to Henry Ford's assembly line, for instance, the production of cars was greatly improved. It *only* took a complete redesign of the production process, but once you pulled that "lever" you were able to produce more cars in less time, with the consequent reduction in production costs.

Engineering advances generate physical levers that we may not be conscious of. For instance, changing the height or angle of an antenna in a cell site improves the quality of the calls we make or the speed at which we can transfer data in our day-to-day mobile communications. Similarly, better software configuration may improve your ability to work on the cloud or on premise. Physical levers require technical expertise that may be costly to acquire or hire, but since modern economies are built around the technological revolution, having at least some general knowledge of what can be achieved can take us very far if we want to be more productive or have more satisfied customers.

Let's consider the design of queues as a final example. Figure 4-2 shows two possible designs: multiple-line, multiple-server on the left and single-line, multiple-server on the right.[1]

This is not the place to even try and delve into the technicalities, but let's just mention that under certain conditions it can be proved that the average waiting time for the design on the left is longer than for the case of a single line. If these conditions are satisfied at your workplace and your objective is to improve general customer satisfaction (measured by the time they spend waiting in line), you can redesign your queues and you may meet your goals.

---

1 In this setting *server* is the person or machine responsible for serving each customer (like a cashier) and not a computer server.

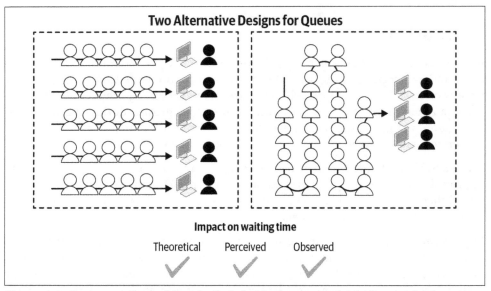

Figure 4-2. Queues as physical levers: the left side shows a multiple-line, multiple-server design; by moving to a single-line, multiple-server design (shown on the right), we may impact waiting time, so this change is a lever we can use when we want to have an effect on customer satisfaction

Physical and Psychological Levers in Waiting Lines

In Figure 4-2 I also claim that the *perception* of waiting times may also be positively affected by switching to the design on the right, but this would take us to the terrain of human levers where psychological laws operate. We will address this topic shortly, but you can check out "Why Waiting Is Torture" (*https://oreil.ly/s2Pwf*) by Alex Stone for some evidence on the psychology of waiting in line.

Human Levers

Just as the design and use of physical levers requires considerable technical expertise, human levers require a thorough understanding of how humans behave. Humans, as opposed to materials, have a very specific set of difficulties of their own. Let's discuss the most important briefly.

The most obvious one is that we can't force others to behave the way we want: we have to *incentivize* them. You can't force potential customers to buy your products, or your employees to work more or be more productive: you need to create the conditions that will lead them to act in ways that are favorable to your objectives out of their own self-interest.

Moreover, we are *heterogenous and diverse* beings: even identical twins that share all of their genetic material behave in different ways. We also have a sense of *agency*: we have intentions and these vary from individual to individual and throughout our lifetimes.

To add one more layer of complexity, we are *social animals*, and our behavior may vary drastically if we make choices surrounded by people or alone. We also learn from experience, a process common for toddlers, the elderly, and everyone in between. Finally, we make errors: we may regret some of our previous decisions, but these may not be easily predictable.

Why Do We Behave the Way We Do?

I will set out on an ambitious agenda and try to condense why humans behave the way we do into three categories that I believe cover a majority of the reasons behind our behavior. I was trained as an economist, so you may see a bias in this enterprise, but hopefully other social scientists won't disagree that much.

I will claim that most of our behavior is driven by our *preferences* or values, our *expectations*, and the *restrictions* we face. These map neatly to the economists' portrait of a rational being, but rationality has little to do with this characterization.[2]

Think about why you bought this book: my guess is that you wanted to learn about AI and how to use it to make better decisions, but since you weren't sure of the quality of the material, you took a leap of faith and hoped for the best. Nonetheless, you could be doing anything else right now: you could be reading some other book, technical or not; watching a movie; sleeping; or spending time with your loved ones. You must have valued reading this book (or at least expected this to be the case). At the same time, you were able to afford the book and have the time to read it, two of the most basic restrictions we generally face.

Does this generalize to any other choices? I believe it does with most choices we make, if not all. In a sense, the claim is almost tautological: ask anyone why they just behaved as they did, and they could easily say "because I wanted to."

Now, preferences come in at least two flavors: we have individual and social preferences, and this distinction allows us to account for the differences in choices when we are surrounded by others and when we are alone.

We will now discuss each of these in detail.

---

2 Rationality has to do with the consistency of choices, which I will not use or claim at all.

Levers from Restrictions

Let's start with the pricing lever, arguably one of the most common actions we take to achieve the specific business objective of increasing our revenues. It is one of our favorite levers, since it directly affects our revenues—price (P) times sales (Q), or $P \times Q$.

Interestingly, revenues depend on price in a way that makes the choice to pull the lever not obvious at all. The difficulty comes from what economists call the "Law of Demand": when we *increase* our price, our sales generally *fall*. Since sales depend on the price we charge, revenues would better be expressed as $P \times Q(P)$ to make clear that our choice of the pricing lever has two effects on our revenues: a positive direct effect coming from the first term, and a negative indirect effect from the latter term. The overall effect depends on the sensitivity of demand to changes in price.

The Law of Demand

Figure 4-3 shows how demand (Q) (horizontal axis) varies with price (vertical axis). Don't be confused by the choice of the axes: for historical reasons, this is how economists depict a demand function, even though prices—our lever—would most naturally be depicted on the horizontal axis. A better term is *inverted* demand function, but the term never stuck.

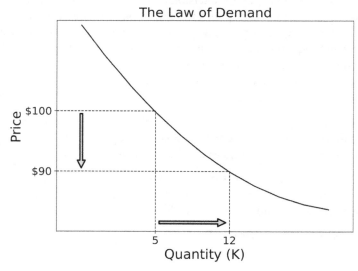

Figure 4-3. *Purchases fall as we increase prices*

The important thing to recognize is that as the price falls, consumers purchase more of our product. Notice that reducing the price $10 from $100 to $90 generates an increase in purchases of 7,000 units. Compare this with Case A in Figure 4-4, where in order to generate the same increase in volume we just need a reduction in price of $1.

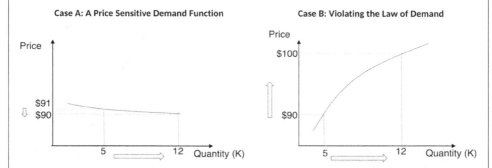

Figure 4-4. Case A shows a relatively price-sensitive demand function, while case B depicts a demand function that violates the Law of Demand

The difference in the two is what economists call the *price elasticity* or our customers' price sensitivity. There are many determinants of price elasticity, most importantly the availability of close alternatives or substitutes and the proportion of our income devoted to the consumption of a specific product.

The right panel in Figure 4-4 shows the case of a product for which the demand increases as prices *increase*, thereby violating the Law of Demand. Are there real-life examples of Giffen goods, as economists refer to them? Take the case of fine wine or jewelry or any other premium goods. For some people their demand will increase when the price is higher, as this may signal better quality or premium status. Say that such a wine and customers exist: if we now decrease the price of each bottle, will they consume *less*? If what they value is the price, this may well be the case, but if they like the wine, irrespective of the price, it is unlikely.

Figure 4-5 shows a somewhat standard relationship between revenues and our pricing lever. It should now be clear that if we want to pull the pricing lever, we'd better know if we're to the right or to the left of the vertical line: our company will be better off if we increase prices in range **A**, since a price increase generates *higher* revenues. The opposite will happen in range **B**. The math notwithstanding, the intuition should be clear: if our customers are not too price sensitive, an increase in price, say by one dollar, will *decrease* demand less than proportionally, thereby generating an overall positive impact on our revenues. This sort of calibration is standard when we are doing price and revenue optimization, one area where prescriptive analysis has been most successful and which will be revisited in Chapter 7.

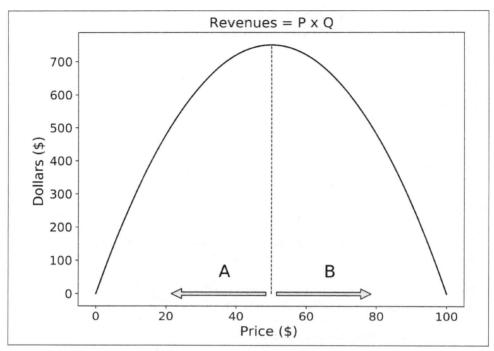

Figure 4-5. How revenues change with our pricing choices

I hope this example convinces you that the choice of the pricing lever is far from obvious, but in my opinion it's one of the most interesting and successful cases of prescriptive analysis. If we are considering giving away discounts, it'd better be the case that demand increases proportionally faster than the falling price.[3] Otherwise, we should look for other levers.

---

3 Unless you want to capture market share that would supposedly increase long-term revenues, but this is a different business objective. In this case you may consider operating at a loss in the short term, but you must then optimize the net present value for your longer-term profits.

Pricing Levers as Restrictions

You may wonder why I chose to classify the pricing lever as a constraint. One important reason why customers generally follow the Law of Demand—the negative relationship between purchases and prices—is that by changing prices we affect their budget constraints.

Interestingly, this effect operates not only with our current customers but also with prospective ones that haven't started buying since current prices may be too high.

But is this always true? The discussion of Case B in Figure 4-5 notwithstanding, most of the time we follow the Law of Demand, so most people just take it for granted.

Time restrictions

It is not a coincidence that two of the main restrictions we face are time and money. We have already discussed the budget constraint, but what about time constraints? Do companies leverage time restrictions as they do with budget constraints?

Consider digital banking. I don't know about you, but most people I know can't stand going to a branch, as it feels like a waste of our valuable time. One of the best cases for a better user experience is that we relax our customers' time constraints, and give them back some time they can devote to other activities.

If you're not convinced by the banking example, ask yourself the following question: is there something you would engage in if the time required was reduced? Imagine that by cutting in half the time in the gym, say from 60 to 30 minutes, you would get the same results. All of those infomercials that promise the perfect abs in just 10 minutes a day are pulling this lever. People value their time as much as their money, because as they say, "time is money."

Levers That Affect Our Preferences

We will now consider some of the different determinants of why we like and value what we do. As we will see, *all* of them are actionable and are constantly used by companies around the world.

Genetics

How much of our behavior is determined by our genetic makeup and how much by our social upbringing? This *Nature versus Nurture* debate is one of the most important and controversial in the social and behavioral sciences, since it is very difficult to disentangle their relative importance empirically (Figure 4-6). For instance, if you enjoy a glass of red wine like your parents, is it because of your genes? Could it be that you were raised watching them enjoying red wine which itself created a positive, but *social*, effect on you?

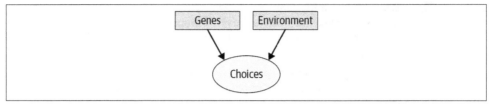

Figure 4-6. Genes and environment both shape our preferences and choices

Let's take the most accepted view that both genes and the environment matter, and that some behavior is most likely to arise when certain genes are exposed to certain environments. We can now ask ourselves if we could leverage this knowledge to attain our business objectives.

At the outset it seems clear that we cannot change our customers' DNA, but in the foreseeable future it is possible that with further advances in behavioral genetics, we'll eventually have a thorough understanding of how to leverage the exposure of specific customers to certain environments.[4] Many stores already do some *very* basic and crude genetic leveraging by changing the aromas present in the store when we are buying. But imagine the case of genetic profiling: a person enters a store, we have knowledge of some genetic markers that matter for our product, and offer a complete sensory experience that makes them more likely to purchase.

I'll leave this here, but keep in mind that this topic raises all sorts of ethical issues. I'll have time to discuss some of these later in the book.

Individual and social learning

The truth about our choices is that many times we don't know what we want or what we like, in contrast with the view of rational and consistent choices put forward by most decision theorists and economists. Some people are more prone to trying new things and exploring the variety and diversity of their tastes. At the other end of the spectrum, other people have had terrible experiences when trying new things and just stick to the same dietary routine they already know and feel comfortable with.

In any case, the fact that preferences are not fixed and consistent, and that most of us like to try new things to at least some degree, should help us find levers to achieve some of our business objectives. This is especially true when a company launches a new product: since customers are reluctant to pay for something they haven't tried,

---

4 See Carolyn Abraham's article, "Why your DNA is a gold mine for marketers" (*https://oreil.ly/w9f0o*) in *The Globe and Mail* for an example.

the company typically gives out free samples. This reduces the real and perceived cost of trying the product and is done in the hope that the customer will be willing to pay the full price next time.

While this applies to both individual and social learning, in the case of the latter we have a second possible lever by way of influential people. In the current age of digital social networks, companies commonly use influencers to help others try their products without the need to give them away for free.

Social reasons: strategic effects

Imagine we see something like the behavior in Figure 4-7: here, a newcomer to the group brings a new idea or behavior. She first convinces one member of the group, who starts behaving similarly. And then another one, and then several others.

How can we leverage this type of social effect to achieve our business objectives? In the previous section we sketched one possible reason—social learning—and discussed two possible levers (price and influential people).

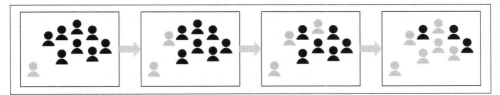

Figure 4-7. Social contagion

But it could also be that there are strategic effects that explain these dynamics. Think about two-sided platforms like Airbnb, Uber, WhatsApp, Facebook, Google, operating systems like iOS or Windows, etc.[5] Going back to Figure 4-7, imagine that one person in our group of friends comes from Europe telling us that they are using the latest messaging app. At first, only her best friend downloads it to try it and, of course, to chat with her. But now two other people try it, because, yes, they want to know what their friends keep talking about! The more people that join, the higher our incentives to join: this is the first type of network effect that operates in two-sided networks.

The second type has to do with the other side of the network. Think about Uber: if more drivers join, it is easier for passengers to find rides, so now more customers join. But the larger demand also makes joining more profitable for the drivers: you can now see why two-sided platforms generate these huge positive feedback loops.

---

5 We also talked about two-sided platforms in Chapter 3.

It is common to refer to these as "strategic effects," since our behavior depends on the choices of others, and vice versa.

Can we use this as a lever to attain our business objectives? Most certainly: one of the most popular levers for two-sided markets is to subsidize the side of the market that is *most* price sensitive by way of discounts or lower fees. This will generate the two positive feedback loops we just described, and by choosing the most price sensitive side we reduce the cost of the lever. For instance, Uber subsidizes the value of each trip by dropping prices for the customers, and Google gives away the use of their search engine, but auctions ad space to the other side.

Social reasons: conformity and peer effects

Many times we change our behavior in response to changes in our social network just because we want to belong. This desire to conform as an explanation for human behavior has its own set of difficulties, most easily illustrated with the case of influencers.

Why would we buy a swimsuit worn by Selena Gomez or Cristiano Ronaldo on Instagram? It could be that we *learned* that it actually looks good on us only after seeing it on them, instead of appealing to a need or desire to *belong*.

Note that conformity may arise from strategic effects: peer and group pressure create a burden on me, so I may find it in my best interest to do what everyone else is doing. The same reasoning applies to my friends and peers, giving rise to what is sometimes referred to as herding behavior.

To sum up, this discussion isn't purely academic: it affects and enlarges our set of levers, especially with certain demographic groups such as teenagers. It may not be that effective with other demographic groups, or at least I haven't seen credible empirical evidence showing that we should care about it in regards to other groups.

As a final note, let's discuss the case of corporate culture, a common use case where conformity might play an important role. A positive culture will make employees happier and more productive, and a negative culture can produce really bad outcomes such as robbery, corruption, and the like. Precisely because we think it matters, it is generally the CEO's and chief human resources officer's task to find ways to create and grow a favorable corporate culture. The desire to conform is but one of the reasons why new cultures arise, so one lever is to find some people who could serve as corporate influencers. Who better than the CEO and their whole executive committee?

Framing effects

Let's now move to the terrain of behavioral economics, the systematic study of "irrational" or "inconsistent" behavior. We'll see that there is a lot of consistency in our inconsistent behavior that can be used to achieve our objectives.

Suppose that given a choice between your product and your competitors', your average customer chooses yours in some circumstances and your competitors' in others. This inconsistency of choice is troubling, since it suggests that nothing intrinsic about your product (or your competitor's) explains the choice, but rather, that something external like the decision context may be the cause of the final outcome.

Consider Figure 4-8 where three alternative TVs are presented with respect to two different attributes, size and price. The problem here is that these attributes compete against each other: I prefer a larger TV, but unfortunately it comes at a cost, so I have to trade-off one for the other. Brand A has the smallest screen and it's thus also the cheapest model. Brand B isn't too different from A (especially when compared with C), and finally C is the best in terms of size, but you'd have to pay some extra dollars to get it. Which one would you choose?

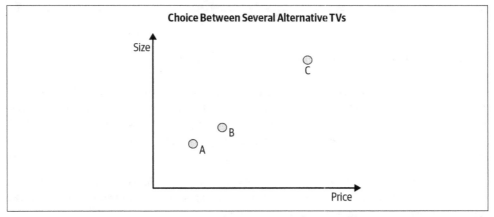

Figure 4-8. Framing effects: each circle denotes a specific type of TV, varying in terms of their price and screen size

If you're like most people, then you would've chosen B. It appears to be the reasonable choice in terms of the two attributes, especially since C is considerably more expensive. Marketers have been studying these effects for a long time, so they usually pull the framing lever to direct our choices to whatever they want to sell. Let me repeat what I just said to make the point clear: they want to sell alternative B from the outset, and to do so they decide to pull a "framing lever." They carefully select the two alternatives they want to display so that we "naturally" choose B.

Consider Figure 4-9 now, and imagine your objective is to buy a new laptop where we only care about two attributes: the amount of memory (RAM) and the speed of the processor (CPU). Case A shows two alternatives that clearly trade-off both attributes: you either have a lot of memory but low CPU (A) or vice versa (B). This type of choice makes us pretty uncomfortable since there is no clear winner with respect to all attributes we care about, and life is so much easier when we don't have to make sacrifices.

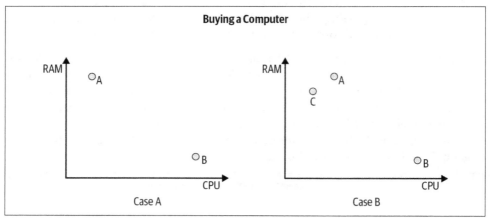

Figure 4-9. Buying a computer: another case of framing effects

Wouldn't it be nice if we could find a reason to choose one or the other? This takes us to Case B, where our retailer now presents a third alternative that is clearly dominated by laptop A (C has less memory *and* computing power). Why would he do that? Alternative C acts as a reference point that helps us find undisputed arguments to now choose A.

Note that the lever here is *the way we present or frame the choice situation*. Compared to other levers that negatively impact our revenues (like discounts), this sounds like an almost free way to increase our sales.

Loss aversion

Our final example of levers that may affect your customers' preferences is known as *loss aversion*. As the name suggests, the idea is that the worth of something changes if we own it or not, or more interestingly, if the choice situation is framed as a loss. Figure 4-10 shows how much we value money by way of a utility function. These functions translate units of the things we own and care about into units of utility or satisfaction, and are commonly used to analyze decisions (see Chapter 6).

The horizontal axis shows our net holdings of money (if negative, we lose) and the solid line shows how much we value each of these states. The important thing to notice is that the function is asymmetric with respect to gains or losses: gaining an extra 25 cents is valued less (in absolute value) than losing a comparable amount. Loss aversion really captures this idea: our brain may be hardwired to make us more sensitive to loses than to gains.

Is this something we can use to attain our objective? It may not come as a surprise by now, but yes, the way you communicate with your customers can make a difference. What the theory of loss aversion suggests—again, backed by tons of experimental evidence—is that framing choices as losses can make a difference.

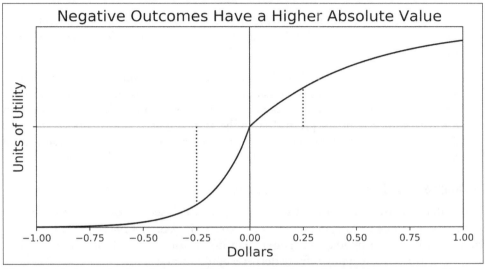

Figure 4-10. Loss aversion: winning 25 cents is judged as a lesser option compared to losing the same amount

Suppose you want to sell the latest version of your product. If you give some credit to this theory, you may try A/B testing something like these two alternative messages:

Alternative A
    "Buy our amazing new product!"

Alternative B
    "Don't miss the opportunity to buy our new product! It's a once-in-a-lifetime opportunity!"

Since alternative B frames the communication as a loss, we should expect to have a higher conversion rate on it relative to A. This may sound crazy, but since testing is *relatively* cheap, why not try it? Recall that our aim is to sell more without having to give out our products at a discount.

Levers That Change Your Expectations

We've now covered preferences and restrictions. Preferences guide our choices and restrictions force us to choose between competing alternatives. What role do *expectations* have, then?

Most of our decisions are made without us knowing the outcome of our choices. Should you date or marry that person? Should you buy coffee or tea? Should you accept that job? If you think about it, all of these choices are made under conditions of *uncertainty*.

Our brain is a powerful pattern-recognition machine that allows us to make relatively good predictions many times. But how do we do it? Do we have the laws of probability hardwired into our DNA?

The work of psychologist and Nobel Prize in Economics-winner Daniel Kahneman and his coauthor, the late Amos Tversky (and their many students and coauthors), has taught us that our brain simplifies many of the computations needed to survive in a world where uncertainty is queen. Two of the most important shortcuts that we take are the availability and representativeness heuristics. We will see that these can be leveraged to attain our business objectives.

The availability and representativeness heuristics

Recall that heuristics are shortcuts or approximations used to solve computationally hard problems like making decisions under uncertainty. Sometimes quick and dirty—though vaguely approximate—is better than no decision at all. That's probably how our brain evolved into a powerful pattern-recognition machine.

Quantifying beliefs requires gathering evidence, which is generally costly (as you've probably witnessed). With the availability heuristic we simplify this process by taking whatever evidence is most readily available and using it to approximate likely scenarios. With representativeness we use whatever evidence we have, even if scant, and extrapolate it. Note that these are shortcuts: if we had more time and resources, we could have collected more and better evidence to form our beliefs.

Let's use this knowledge to understand how we can improve our advertising efficiency. The difficulty with advertising is that it doesn't necessarily have an immediate impact, but it may affect our brand awareness. When a potential customer actually wants to make a purchase, the availability heuristic may bias his recall toward our brand if our campaigns were successful.

What about representativeness? If your first product was really good, your customers may extrapolate this belief, making them more willing to purchase a second one. Or, thinking about issues of corporate governance: if you have created a reputation for disrespecting the most basic ethical standards, customers may extrapolate that to the quality of your product. Choice heuristics abound, so we should use them in our favor (and be extra careful not to have them work against us).

Revisiting Our Use Cases

The last pages presented a lot of material. My goal was to point out different sources of inspiration to find levers to achieve our business objectives. This will be more apparent now when we revisit the use cases from Chapter 3.

Customer Churn

Just as a reminder, we first framed the business question as a prescriptive one, and we now want to start looking for levers to achieve such an objective. In Chapter 3 we concluded that our aim is not to minimize churn, but rather to maximize profits from our retention campaign. In this case it may be optimal to let some customers go if it's just too costly to keep them loyal to our brand.

What actions can we take to achieve this business objective? Think about it: why on earth would someone want to be our customer instead of our competitors'? Let's go back to basics.

Customers generally want three things from our companies: good-enough quality products, affordable prices, and good customer service if they need support. Moreover, they'll likely be willing to trade-off one or more of these, at least to some degree. To leverage this knowledge we must move to the terrain of preferences, restrictions, and expectations.

With a price discount we may be willing to sacrifice short-term profitability if the long-term impact is positive and incremental. But this is not the only lever that we have. We can also create the perception that switching is costly by creating a loyalty program or highlighting some of the least favorable attributes of our competitors' products. The fact that they know our product (rather than our competitors') should help us design levers that take advantage of loss aversion or availability heuristics (recall the adage "better the devil you know than the devil you don't").

On a final note, what about some of the consistently inconsistent behavior we mentioned previously? Most economists believe these will work as levers only temporarily, and eventually your customers (or some competitor) will realize that they're being framed. You may exploit these short-term rents, but be careful if you're considering making them an integral part of your business model.

Cross-Selling

In cross-selling we are looking for the next-best offer for each of our customers so that we maximize their customer lifetime value. In this sense, our main lever is to offer, or not, each of our products to each of our customers. Note that we may want to include the "no-offer" option as a lever, since we may lose customers by making undesired offers, thereby reducing their lifetime values.

This said, you can use some of the techniques we have described as second-order levers, that is, as levers used to accomplish the actual cross-sale. For instance, the way we communicate and frame our offers can always be used in our favor.

Capital Expenditure (CAPEX) Optimization

While the immediate levers have already been spelled out for use (spend or not spend), the intervening factors that map spend to consequences are far from well understood.

For instance, CAPEX is one of those cases where physical levers can play a significant role (think of investing in cost-saving technologies). But we may also think of human levers that affect our revenues (invest in new technologies that make our products more desirable, even if the cost is higher, until we're able to achieve some scale economies).

This class of problems is so broad that the search for levers must be done on a case-by-case basis.

Store Locations

Similar to the problem of CAPEX optimization, our immediate lever is to open stores or not (or even close some) at different locations, and again, it's everything but clear why these actions impact our business objectives. At this level it's almost magical that we just decide to open a store and our bank accounts start collecting more profits.

The most important intervening lever is our ability to capture some share of demand (price or quantity) or reduce our costs (that also vary spatially). We will have more time in later chapters to discuss these issues in depth.

Who Should I Hire?

Recall that our objective here is to maximize the incremental returns from hiring. For this we must have a good understanding of the employees' impact on our business, which, as we'll see in detail later in Chapter 7, is not obvious at all. But assuming we have this piece of information, our decision is then to hire or not, and at what cost. Again, we have a binary lever (hire or not) and one that we can fine-tune more granularly (the salary, benefits, emotional salary, work environment, and all other levers used by recruiters).

Delinquency Rates

The business problem is to maximize the ROI from lending resources to one customer. As such, the three natural levers for this use case are the size of the loan (zero inclusive), the time horizon or maturity, and the interest rate if regulation permits. At this point we can forget about the complexity of optimizing all three: we must first start understanding the menu of levers we have at our disposal.

But we can be way more creative and test behavioral levers. What if we print children's photos on credit cards? Will that make our customers *more* likely to pay their debts on time? Or talking about communication strategies, can we *nudge* better payment behavior just by sending an SMS with a happy emoticon? Again, testing is relatively cheap: we just need a working hypothesis, the ability to think out of the box, and stakeholders' buy-in to find less costly levers.

Stock Optimization

At the most basic level, we want to leverage how many units of each item we should have in stock. Levers then are just a number, which could be positive (we need to have more stock), zero (current amount is just right), or even negative (move some of these items to other stores since we will never be able to sell them at this location). This is not to say that there aren't other physical levers that can be pulled. For example, think about reducing our first order storage and transportation costs (think Amazon).

Store Staffing

The choice of levers in this problem is again restricted by physical and operational constraints. For instance, is it operationally feasible to make specific staffing decisions for any given hour of any given day? What about every half hour? Recall that we should have the right number of salespeople in each store in order to maximize our profits or customer satisfaction. But this depends on how many customers we have at any given moment, so depending on the granularity, we may always be under- or overstaffed.

If we are willing to think out of the box, we might even consider relaxing these operational constraints by "Uberizing" our staff: hire people only when demand is high enough.

Key Takeaways

- *Once we define a business objective, we must consider whether it's actionable*: most times our problems are actionable, but we may have to think out of the box.

- *The problem of choosing levers is one of causality*: we want to make decisions that impact our business objectives, so there must be a causal relation from levers to consequences.

- *The two main types of levers are physical or human*: physical levers are usually exploited by taking advantage of technological advances. Human levers involve a deep understanding of why our customers, employees, or any other human being involved with our production process behave the way they do.

- *To understand the relation between actions and consequences, we must construct hypotheses*: most times we don't need to rediscover the wheel, as there's plenty of knowledge out there about how things work or how humans behave. I provided a very quick and incomplete overview of some findings that I've found useful to think about these problems.

- *Hypotheses often fail, but we should embrace the learning process*: many times we start with a theory about causes and consequences only to see it fail during testing. That's fine. It's part of the process. Embrace it and guarantee your team learns from these failures.

Further Reading

Physical levers are problem-specific, so my suggestion here is to ask your more technical colleagues for some reading suggestions that can help you gain at last some general knowledge of what can be achieved or not.

The human levers discussed here have been thoroughly studied by social scientists, including economists, psychologists, and sociologists. I'd recommend starting with an introductory textbook on microeconomics since most of our business decisions have some economic foundation. I personally enjoy how David Kreps, professor at Stanford Graduate School of Business, explains these very technical topics to the more general public.

Books on behavioral economics will give you some extra background on less-than-rational choices. One of my favorites is Dan Ariely's *Predictably Irrational* (Harper Perennial), but a safe choice will also be Daniel Kahneman's *Thinking, Fast and Slow* (Farrar, Straus and Giroux). I certainly recommend Ariely's book as he provides many examples of levers we might pull to improve our business that we would never have expected to work. If you want to learn to think out of the box, this is the type of literature I'd recommend.

Gary Becker and Kevin Murphy's *Social Economics* (Belknap Press) is still a good reference on choice under the social umbrella, but another classic, full of intuition, is Thomas Schelling's *Micromotives and Macrobehavior* (W. W. Norton & Company). A highly technical and encyclopedic treatment of the economics of social networks can be found in Matthew Jacksons's *Social and Economic Networks* (Princeton University Press).

Finally, strategic behavior and game theory are topics of their own. You can start with an introductory textbook that is less technical but provides a lot of intuition. Avinash Dixit and Barry Nalebuff's *The Art of Strategy* (W. W. Norton & Company) or an introductory textbook by Ken Binmore, such as *Fun and Games* (D. C. Heath & Co.) or *Playing for Real* (Oxford University Press), might do the job.

From Actions to Consequences: Learning How to Simplify

We've mentioned several times already that we impact our business objectives by making decisions that may or may not have the consequences we expect. But how do we know what to expect? And how did we decide that this set of actions was worth the time spent experimenting in the first place? In this chapter, we will learn one of the most important analytical skills: the ability to simplify the world and solve restricted problems. This will hopefully take us a bit closer to analytical enlightenment.

Revisiting the graphic shown in Figure 5-1, we will now put our focus on the connection between actions and consequences, in a process mediated by causation. This is where we put into practice our outstanding ability to create theories about how the world works. But since the world is complex, we'd better learn how to simplify.

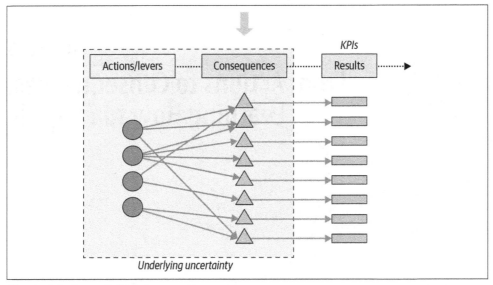

Figure 5-1. From actions to consequences

Why Do We Need to Simplify?

Imagine that we start with a project to solve a new business question. Following the recommendation in Chapter 3, we set up the problem by starting with the question and moving backward. We first identify some drivers behind our objectives and then come up with theories that map our levers to these consequences.

It is not uncommon, however, that for any given consequence we're going after, there are many levers we can pull. Each of these levers can create a plethora of consequences themselves, hence our need to simplify (Figure 5-1).

Say we want to increase our revenues. In Chapter 4 we showed that depending on our current price, it could make sense to *increase* or *decrease* our prices. But you may very well just send a daily email to your customers saying that you appreciate their business and wish them a nice day. Or ask your community manager to post on the company's Instagram account some cool pictures of cool people using your product. Or wait until they reach out to you at your customer service center and then try to do some cross-selling. Or if you feel really adventurous, you can place an ad on a Bolivian webpage hoping that some of your potential customers are planning the next trip to that beautiful country. Literally, there are unlimited options. The world is complex, and some of these may work with some of your customers, and some may not. That's why we need to simplify.

First- and Second-Order Effects

One good practice is to start considering only *first-order effects* with the objective of getting the "sign" or "direction" of the effect right. At this point we may not care about the *second-order effects* that affect the curvature of the outcome.

But where do these effects come from? Since I'm suggesting you start with first-order effects, I'd better tell you where to look for them. As the following sidebar shows, we need a theory of how consequences arise from actions. The theory may be in mathematical form (where you end up mapping actions to consequences) or rather more story-based, where you just hypothesize that if you pull lever x then consequence y follows. With theories written in math you can derive first- and second-order effects. With less formal theories, you'll just have to rank alternative levers with the benefit of hindsight.

Taylor Series Approximation

It is worth understanding where the expression of first- and second-order effects comes from. Given a "nice" function $f(x)$ that depends on x, and a value a, the second-order Taylor series approximation (at a) is

$$f(x) = f(a) + f'(a)(x - a) + \frac{1}{2}f''(a)(x - a)^2 + Res$$

where the usual prime notation denotes the first (f') and second (f'') derivatives of f evaluated at the value a, and Res is the residual from the approximation that depends on higher-order terms.

Under this approximation, the first-order effect is the second term on the righthand side, capturing the sign or direction of moving away from a. Note also that it depends *linearly* on $(x - a)$. The second-order effect (third term on the righthand side) captures the curvature, involves the second derivative, and depends nonlinearly on how much we deviate from a, $(x - a)$.

The function $f()$ is our theory that maps actions x to consequences, and you can think of $f(a)$ as the current state of the world. We want to evaluate the counterfactual state $f(b)$ and see what consequence arises.

Part of the complexity we face comes from the fact that we live in a highly nonlinear world. Think of the relationship we saw between revenues and price (Figure 4-5). Why would we need to start with an inverted U shape? It is simpler if we just start with a linear and *increasing* function from price to revenues (at least we know that there is a direct and positive first-order price effect).

Paraphrasing Einstein, models should be as simple as possible, but not simpler. This means that we should start with the simplest theory that fits some data or our prior knowledge of the problem, even if partially, and then, if necessary, iterate and create more complex models.

A good rule of thumb of when to stop can be found by way of standard cost-benefit analysis: stop when the incremental cost of another iteration exceeds the benefits you expect to get from it. If you need to invest one month of your team's time in order to get out the next version of your model but this only provides a marginal increase in your company's revenues, it feels like you shouldn't have asked for that iteration in the first place.

Exercising Our Analytical Muscle: Welcome Fermi

Enrico Fermi was an Italian physicist and member of the Manhattan Project. He played an important role in the development of nuclear physics and the atomic bomb. Among non-physicists like myself, he is most famous for popularizing a method to approximate solutions for very hard and large problems. As an example, let's take a problem named after him (but for somewhat different reasons), the so-called *Fermi Paradox*: why haven't we found evidence that there is life in the universe when *the numbers suggest* that it's not such a low probability event? Well, we need to come up with these "numbers" to verify the apparent paradoxical nature of this statement, and that's where the Fermi method comes to our aid.[1]

Fermi problems start by decomposing a complex problem into simpler subproblems. Our intent is not to come up with a hard, precise number, but just to have the correct *order of magnitude*: we may not care if the answer is 115,000 or 897,000, as long as we know that it is in the order of the hundreds of thousands (10^5).[2]

I love Fermi problems as they force us to exercise our analytical muscle by demanding we make simplifying assumptions and rough approximations in order to provide an answer. Let's see how this works in practice.

How Many Tennis Balls Fit the Floor of This Rectangular Room?

Many tech companies are famous for asking Fermi problems during their interviewing process, so let's start with the archetypical example. Suppose you are standing in a rectangular room of length (L) and width (W) as in Figure 5-2. How many tennis balls can we fit on the floor of this room? What about inside the whole room?

---

1 Think about it: how could we even approximate this probability? We need to essentially count the number of Earth-like planets in the surrounding universe.

2 You can decide if this is too rough an approximation for your problem and adjust accordingly.

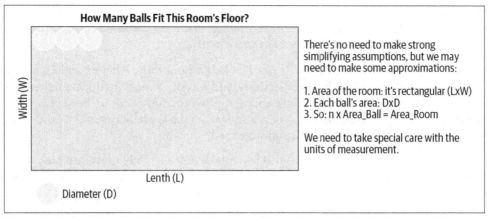

Figure 5-2. Applying the Fermi method to a problem of fitting balls in the room's floor

Consider first the number of balls that we can fit in that first row on the top left of Figure 5-2. Since you only care about that first row, we can neglect the width (W) as it doesn't provide much information. The number of balls is approximated by $n_b = L/D$, or better, the greatest integer less than or equal to that number[3]: if the room is 10 meters long, and each ball has a diameter of 25 centimeters, then we can fit 40 balls in each row. How many of these rows can we fit now on the whole floor? Using the same logic, we see that the answer must be $n_r = W/D$.

Multiplying these two numbers, we get that the total number of balls we can fit on the entire floor is given by $N = n_b \times n_r = \frac{L \times W}{D^2}$, or whatever nice integer from below this gives us. To get a numerical answer, you need to come up with approximations for each of these values. For instance, I believe that a tennis ball is between 5 and 9 centimeters in diameter (between 1.9 and 3.5 inches), so let's take the midpoint of 7cm. Also, my room is approximately 4m long and 2m wide, so I should be able to fit something like 1,600 tennis balls. We know that this is not exactly correct, but we want it to be in the ballpark. Approximation and simplification are at the core of Fermi problems.

But let's focus on the technique itself to see what we did. We started with a problem and decomposed it into two relatively simpler problems. The first one is how many balls we can fit in each row; the second is how many rows of balls we can fit in the rectangular room. If we are able to solve these, we can then address the original, much harder problem. You probably noticed that we could've found the solution in one step using the corresponding areas, but that's just a bit more complicated (see the

---

3 We do this because we can only have an integer number of balls, of course.

explanation in Figure 5-2). If you now want to get the number of balls you can fit inside the whole room, you have a third problem that is on the same complexity scale as the first two (how many slices of balls you can fit).

Also, notice that the area of a ball is not $D \times D$ but rather πr^2, where r and D are the radius and diameter of the ball, respectively, and π is the almost mythical mathematical constant. But since we care only about approximation, we don't need this exact formula: we can just assume each ball is a square, as the space between the borders of each of these squares and the ball cannot be used.[4]

Let's see what simplification buys for us by considering a slightly more complex room like the one in Figure 5-3. How can we solve this problem now? A first-order approximation would treat everything as a rectangle, so the problem is essentially the same, with the difference that instead of L in the formulas you would use $L + k$. A linear world is easy to handle.

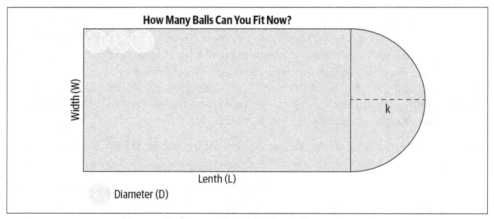

Figure 5-3. Same problem as before, with a slightly different-shaped room

You may feel uncomfortable with this rough approximation, though, so you can now provide a lower bound to your estimate by approximating the half-circle on the right with a triangle, or if you really want, you can try the area of a circle to provide a much better approximation. Was it really worth the extra effort? You can play with numbers and see if it was.

I know this feels like an artificial problem, but remember that we're just warming up in our first Fermi workout.

---

4 It would be different if you were allowed to cut the balls in pieces to fill as much space as you can. Then we would need to use that extra empty space.

How Much Would You Charge to Clean Every Window in Mexico City?

The second question has more of a business flavor: how much would you charge to clean *each and every* window in Mexico City? If you've never been there, how could you even provide such a number? Fermi never visited the rest of the universe when he provided an approximate number of habitable or Earth-like planets, either. That's part of the beauty of Fermi problems.

Let's start by setting up the question. We need two numbers: the number of windows in Mexico City and how much to charge per window. We can then multiply them and we are done.

Let's start with our estimate of how much to charge per window. Notice that this itself is a hard question. I would charge differently at night, during the day, or on a weekday, as my opportunity cost depends on the alternatives and how much I value them. But we will simplify and assume that things are constant (this is a linearity assumption).

How Much to Charge per Window

Cost per Window = Time Spent per Window (Hours) × Cost of an Hour

It is useful to check the units to see that everything went well. The units for our answer should be in dollars per window, so:

$$\frac{\$}{Window} = \frac{Hour}{Window} \times \frac{\$}{Hour}$$

How much time do we spend per window? I'll guess that it takes me 30 seconds to finish one. And I will assume that this time is constant. I *know* this is not true from experience: at the beginning I might be very productive and finish in 20 seconds, but then I start getting tired and this rate starts falling. Note that this is a second-order effect as it relates to the *curvature* of the change. On a first pass I will assume away these nuances and just approximate the *average* time to be 30 seconds per window. Since there are 3,600 seconds in an hour, these represent close to 0.008 of an hour (10^{-3}).

What about the cost of an hour? Assuming away all the interesting nuances, I will just say that if I make \$100 a year, then my *average* hourly rate should be close to $100/(52 \times 5 \times 8)$ since there are close to 52 weeks, each with 5 weekdays, each with 8 working hours. Please note how we simplified things out.

This said, we can now estimate the number of windows in Mexico City. Let's consider the case of *residential* windows, and I'll leave the remaining cases for you to solve (see Figure 5-4).

Approximating the Number of Residential Windows in Mexico City

We can apply the same trick as before: multiply and divide by some *meaningful* number so that the product of ratios is now easier to approximate.

$$\text{Windows} = \frac{\text{Windows}}{\text{House}} \times \frac{\text{House}}{\text{Population}} \times \text{Population}$$

If we can estimate each of these ratios, we will be able to provide an approximation. But let's first stop and notice what we've done: this is an equation—a mathematical truth with no ambiguity whatsoever—where we are just multiplying and dividing by some number. Simplification and approximation will make their entrance when we start filling out each of these ratios.

For instance, how many windows do you have in the average house? Say the average house has two rooms and a living room and a kitchen. Do you feel comfortable with an estimate of 10 windows for this average house? That's two and a half windows per room, three for the living room, and two for the kitchen (or some combination of this). I'm sure you have seen flats with more (or less) windows, but remember, we are approximating to the first-order, and my guess is that there are in the order of 10^1 windows per house.

Consider the second ratio now: how many houses per person are there in Mexico City? Don't worry, I don't know either, but let's say that the average house has 4 people, corresponding to a ratio of 0.25 households per capita. Finally, what about the population of Mexico City? I know it's one of the largest cities in the world, so I'm guessing that the population is 20 million inhabitants (recall we only care about the order of magnitude). There we go, our estimate of residential windows is around 50 million (Figure 5-4).

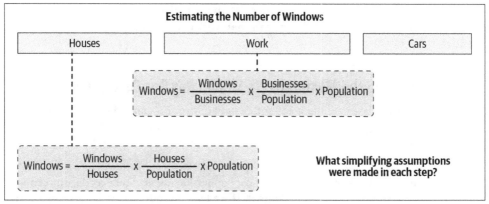

Figure 5-4. How many windows are there in Mexico City?

We've made many rough approximations, since in Fermi problems we only care about orders of magnitude, so we know we may be off with one or several of the numbers we provided. The good thing is that at any point, if we feel we have oversimplified one estimate, we can always go back and improve on it.

I'll let you think about the other nuances of this specific problem, like how many windows there are in restaurants, cars, and schools. But you can expect the method will work analogously. The important thing is that we learn to simplify and be aware of all assumptions made in each step.

Fermi Problems to Make Preliminary Business Cases

The problems in the previous section are great to exercise our analytical muscles, as they force us to make simplifying assumptions and rough approximations, while at the same time being conscious and critical about them. And while I certainly recommend doing a daily Fermi workout, you may wonder if it serves other purposes. We will now see that we can use Fermi-type reasoning to build the logic needed to prepare many business cases in our day-to-day work.

Paying our customers for their contact info

Many companies have terrible customer contact information: we either saved it incorrectly due to human error, it's outdated, or our customers didn't provide the right info from the outset. In any case, their phone numbers or email addresses may not be correct or even usable. This is not only frustrating; it also has a financial impact, since most of our direct marketing campaigns depend on us actually being able to contact our customers.

Your CMO asks you to figure out how much we should pay our customers for their correct contact info. We wouldn't really pay them directly—we could, though—but we can buy some trips to Paris and then let them register for a game or lottery by giving us their contact info, which we can then verify by sending them a code. You get the idea.

Finding the Maximum We Should Pay Our Customers

The trick is to approximate the costs and benefits from this campaign. The cost is the price we are willing to pay to each customer. We can bound this by the *maximum price*, p_M, defined as the point where the campaign breaks even on a per-customer basis.

On the benefits side, the main reason we care about having accurate contact information is because this acts as a sales channel. If we take a random customer we wish to contact, we need to calculate the probability she accepts our offer and the incremental value if it's accepted (ΔCLV). Since we care about quantifying the value of having the right contact info, the probability can be further decomposed into the probability that we have correct contact info (q_c) and the probability of the customer accepting, given that we have the right contact info (q_a). The break-even customer is found when the incremental cost equals the incremental benefit:

$$p_M = q_c \times q_a \times \Delta CLV$$

Let's think about the economics of the problem. The cost for us is whatever we are willing to pay each of our customers, let's call it p_M. But what is the benefit? What do we expect to get back once they give us their contact info? Let's make a rough approximation of the conversion funnel.

Let's assume that out of 100 registered customers, q_c percent gives us the *correct* info; we can make it go to 100% by forcing immediate verification, but right now we don't lose anything by having more generality. Our objective is to make a sale, thereby increasing each customer's value to the company. Let's say that q_a percent of those contacted accept our offer. Finally, what is the impact on our business objectives? I really hope that whatever we are doing increases their value for the company. I'm going to say that, on average, our campaigns generate an incremental long-term value of ΔCLV.

Cool. Now, let's equate the incremental cost and benefit to find a break-even rule. To a first approximation, we should be willing to offer *at most* $p_M = q_c \times q_a \times \Delta CLV$ dollars to each customer. Notice that if we estimate a higher potential gain—as measured by the change in CLV—we can make a better offer too. Also notice that we can, to a first approximation, assume that both relevant probabilities—contact and acceptance—do not depend on the potential gain for each customer; but this assumption is something we may want to revise later on, as it feels too strong. You see the point; we can now just plug in some values that seem reasonable and deliver an answer to our CMO.

Excessive contact attempts increase the probability of churn

Our CMO was quite happy with the answer we gave her since she actually has a budget to buy some trips to Paris and test the campaign. She's now worried that we are going to be so successful updating our customers' contact info that we may end up contacting them too much, thereby increasing the rate of churn.

She asks us now to find a rule that we can follow in order to decide which customers to contact. Let's go back to the incremental cost-benefit analysis. Since we care about the churn probability, let us give that a name, say p_c. If a customer churns we lose her current CLV, assuming she leaves and never buys from us again: CLV_0. Yes, that seems like an extreme assumption, but we are simplifying, and this can be thought of as a *worst-case* scenario. That's the downside. What about the upside? If she doesn't churn and accepts, we can increase her value by $CLV_1 - CLV_0$; let's call that probability q_a.

Basic economic analysis suggests that it's bad business if we call anyone for whom the expected cost is larger than the expected benefit. As before, we can find the break-even customer by equating marginal costs and benefits:

Who Should We Contact?

We can find the marginal customer when the cost from calling that customer equals the benefit, or $p_c \times CLV_0 = q_a \times (\mathrm{CLV}_1 - \mathrm{CLV}_0)$. This can then be rewritten as:

$$p_c^{max} = q_a \frac{\Delta CLV}{CLV_0}$$

We now have a simplified, yet optimal rule for contacting our customers. How could we put it in practice? Consider the simplest case first where we use average churn and acceptance rates as estimates of our probabilities (so we take them as fixed for all customers). To find a break-even customer, we must have at hand customer-level estimates of the percentage incremental value for the offer ($\frac{\Delta CLV}{CLV_0}$) and sort our customers in increasing order. We can then find our break-even customer by equating both sides (Figure 5-5).

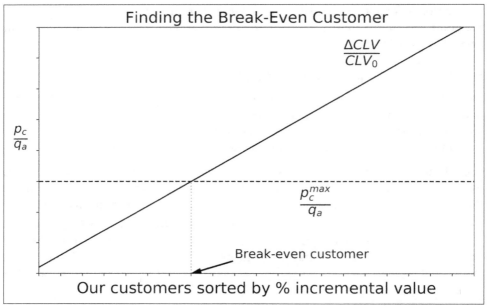

Figure 5-5. *Finding our break-even customer*

Once we are able to solve this simplified scenario, we should consider estimating individual probabilities to have a fully customized optimality rule. Note again that we start by solving a very simple formulation and then move upward on the complexity ladder.

Should you accept the offer from that startup?

Imagine you get an offer to join a promising startup. Because of their lack of liquidity and urge to quickly grow revenues, it's not uncommon for startups to make lower-than-average salary offers, but compensate you with stocks or restricted stock units (RSUs). Essentially, they want you to trade-off short-term liquidity (your salary) with medium- to long-term returns given by your expectation of what they'll be worth in the future.

One difficulty is that RSUs usually have a *vesting period,* a minimum time period you have to work for the company before you can exercise or sell those stocks. For instance, in the United States it's somewhat standard to set this vesting period to four years, so if you're hired today, *and if* you're still employed with the company in four years, you'll have full rights to your RSUs, assuming you can actually sell them.[5] For

---

5 I'm assuming away the possibility that some fraction of your RSUs will vest in regular intervals *before* your vesting period comes to an end, as is usually the case.

this, the company must have gone public (or been acquired) in the interim. Otherwise you'll have to wait until this happens since, simply put, there's no market to trade those stocks.

Basic economic analysis suggests that you should only accept if the offer is better than your current status. Note that other considerations (e.g., reputation of working at one or the other place or emotional salary) are simplified out of the analysis. But how much are you really being offered?

Let's write down some math to get a better idea. Call your current yearly salary w, and to make the trade-off clear, let's say that they offer you just a fraction k of that, say kw. If you keep your current job, you will make w every year. On the contrary, if you accept the offer you know that you'll get kw each year during an *uncertain* period, and you'll also get Sp_S worth of your RSUs if you can exercise your rights to that little part of the company you own. The offer they make includes S stocks, each one valued at p_S dollars.

Should We Accept This Offer?

The cost of accepting any offer is the net present value (NPV) of your current income stream: let's call it I_0. The NPV of the offer will be denoted by I_1, so simplifying away any other concerns you may value—such as the *emotional* salary—you should only accept if $I_1 > I_0$.

In a two-periods problem (today versus tomorrow), the NPV of your current salary is just $I_0 = w_0 + \frac{w_0}{(1+r)}$, where w_0 is your current salary—assumed constant and safe—and r is the discount rate. The same thing happens with your offer, but *tomorrow* you also get the value from the RSUs (Sp_S).

You should accept when:

$$w_0 + \frac{w_0}{(1+r)} < w_1 + \frac{w_1 + Sp_S}{(1+r)}$$

This is a problem where we need to calculate net present values. We cannot simplify away time since there's an opportunity cost of waiting until you own the stocks and can liquidate them. But we can come up with the simplest NPV calculation, one that involves only two periods: today and tomorrow, and we will decide later what "tomorrow" means.

The NPV of staying at the current job is: $I_0 = w + \frac{w}{(1+r)}$, where r is an *appropriate* discount rate. This just says that if *today* we make w, the same salary will be worth somewhat less *tomorrow*: $\frac{w}{(1+r)}$. On the other hand, if we accept the offer, we *expect* to get:

$$I_1 = kw + \frac{kw}{1+r} + \frac{Sp_S}{1+r}$$

Here we have explicitly defined *tomorrow* as the time when you can exercise your RSUs. Now that we have posed the problem, our decision rule is accept if and only if $I_1 > I_0$.

Thanks to this math we can now take it to the next stage and plug in some values: your salary, the alternative, the number of stock units S, and the current price estimate p_S better be part of the offer, but double check the price since there is no market yet and you probably got the info directly from your recruiter.

To make a decision, we must finally plug in a value for the discount rate r and this depends on how many *years* (T) you think it will take the company to go public: if you think it's less than four years, then $T = 4$ (remember the vesting period). If you think it could take longer, then $T > 4$.[6] There is a nice relationship between r and T that will help us come up with an approximate answer.

The trick here is to set up a reference yearly discount rate i and use compounding. The most conservative discount rate is usually given by the inflation rate, so you can use this as a starting place. It then follows that $(1 + i)^T = (1 + r)$, or solving for our unknown discount rate (as a function of the discount rate and the time horizon), $r(i, T) = (1 + i)^T - 1$. We can now compute everything, run some sensitivity analysis changing some of these, and decide if this is worth it for us. For instance, you can find the lowest salary you would accept and then negotiate on the number of RSUs they are offering.

Revisiting the Examples from Chapter 3

Let us revisit our examples and see how much we need to simplify in order to tackle each of these prescriptive problems.

Customer Churn

The difficulty with customer churn is that we have several alternative potential levers, including price discounts; more frequent, nicer, or more strategic communication

---

6 So $T = max\{4, T^*\}$ where T^* is the time you expect it will take for them to go public. Try to inform yourself by reading the news about the company.

with our high-probability customers; and even revamping our product to offer better quality. These work in practice to the extent that the theories we have about the reasons our customers buy from us are valid. And theories rely on simplifying assumptions.

Take the discounts lever, for instance. The hypothesis is that if we drop our prices temporarily, our customers will remain loyal to our company. In this case we can appeal to the basic microeconomics of the Law of Demand, but this is not a law in the physics sense of the word.

To see this, consider the following scenario. We drop the price as proposed by the retention strategy, and since for many customers price signals quality, some of our customers perceive this as a drop in quality and decide to buy from our competitors. To be sure, I'm making two assumptions: price acts as a quality signal for our product, and some customers are willing to pay more to get the quality they expect.

So we have two competing theories about the same lever (temporary discounts) with opposite conclusions: in one we drop our prices and customers stay with us. Under the second one customers leave. What do you think?

If you ask me, I think both are reasonable under certain circumstances, and I've met people who may act as pure discount chasers and those who are willing to pay a high-enough price for quality. But to me, the first-order effect—the one that most likely impacts our *average customer*—comes from the first theory, while the second one would be more of an edge case (affecting the tails of the distribution).

Of course, this is pure theorizing: we now need to take it to the experimental arena and run some well-designed A/B tests. But at least we can start with the most direct lever.

Cross-Selling

What is the next-best offer to be made to our customers if our aim is to maximize their customer lifetime value? Consider the best-case scenario, one of complete personalization: offer the *right* product to the *right* customer at the *right* time and with the *right* communication at the *right* price. Each of these *rights* is a lever for us: products, timing, pricing, and the choice of our customers are all decisions we can make to achieve a successful cross-sale.

But it also suggests that each of these "rights" exist for customers. Take the first natural lever: offering the "right" product. The idea is that given a customer's previous purchases, there is another product that we offer that is right for them. Why is this the case? The truth is that there may not be one, at least on our menu of options.

Another alternative theory is that given the right combination of the other levers (communication, timing, and pricing), *any* customer is likely to buy *any* given product. Some will never do it, no matter what combination of levers we choose, but others may not be as relentless. Thanks to the discussion in Chapter 4 on behavioral economics, we know that communication, pricing and timing may help us frame our customers' decision in satisfactory ways.

You can see that for our levers to work we have identified simplifying assumptions in each of our theories. One common assumption that we haven't discussed is the idea that all customers are the same. This assumption allows us to simplify away differences and focus on our *average* customer. It's fine to start with this assumption, at least to the first-order, but when we go testing our theories, evidence will clearly show us that customers behave in very unexpected ways.

CAPEX Optimization

The problem of capital expenditure (CAPEX) optimization is one where the levers have been decided from the outset (where and how to allocate investment), so we need to come up with a theory of how this affects the business objective we want to optimize. There is no general answer for this, and you should consider it on a case-by-case basis.

I will now propose one possible theory that may be more or less suitable in different scenarios: I will assume that CAPEX affects our revenues $R = P \times Q$, restricting the possible drivers to a pure price effect (P), a pure volume effect (Q), or both.

Price effect

Suppose that by investing more, our company is able to attract customers with a higher willingness to pay. One case where this is plausible is where investment is allocated with the objective of improving the facilities or the user experience. We could then expect that the average ticket our customers are willing to pay depends on the size of the capital allocation $R = P(1 + g_P(x)) \times Q$, where $g_P(x)$ represents mathematically our theory of this direct price effect: this is a growth rate that can be positive or negative depending on this period's capital stock x.

Quantity effect

Alternatively, we may decide to charge the same price, but since we are providing a better-quality product or experience, our sales increase: $R = P \times Q(1 + g_Q(x))$. As before, the function $g_Q(x)$ is our theory that ties capital allocations to our revenues, this time via the volume channel.

We now have two alternative theories of how investment affects our revenues. Needless to say, either one can be valid for our use case, but it could also be the case that none is true, so we would have to go back to the drawing board to come up with

---

better alternatives. For instance, it could be that our investments affects our costs rather than our revenues.

Note that we haven't been specific enough about the actual drivers in our theories, but we have at least suggested some plausible stories. At this level this is the most important simplifying assumption. Once we make assumptions about how to model the actual growth functions in each of the two scenarios, we will have another strong simplifying assumption that we may wish to be critical about.

Store Locations

In the case of finding optimal locations for our stores, the levers have also been decided at the outset, as our problem is deciding where to open new stores. Specifically, our lever is the choice of a location and we therefore need a theory to match locations to business performance as captured by our profits.

The most natural theory is that our revenues will be higher if we open stores where our demand is. There can now be two possible effects: a price effect (targeting higher or lower value segments as measured by the willingness to pay) and a volume or quantity effect. What about costs? Rental and utilities costs will also vary with locations in a systematic way: if many other businesses think like us, there will be higher demand for locations that attract many potential customers (e.g., a shopping mall).

At this level, the main simplifying assumption is to directly map locations to revenues and rental or utilities costs, leaving aside any other indirect impacts. For instance, if we open a new store close enough to an existing one, will there be revenue cannibalization from one store to the other? Most likely we can't safely assume away this effect. The timing of the effects may also be important to consider. If we open a store, how long will it take to generate the stream of profits our theory suggests? What if the competition reacts? You can see that there are many hidden assumptions in what first seemed an innocuous assumption.

Delinquency Rates

This use case is a good place to present the usual trade-offs one may face when simplifying a problem. In general, we get tractability at the expense of providing a less realistic picture of the problem. Our hope is that by focusing on first- and second-order effects, we can solve a problem that can still inform our business decisions in a meaningful way.

In the case of delinquency rates, we have already mentioned that there are at least three natural levers: the size of the loan L, the interest rate i, and the maturity or length of the loan T.

Let's start by posing the most general problem. Our objective is to maximize profits, which can always be represented as the difference between revenues and costs, each of which may depend on each of our levers.

$$\text{Profits}(L, i, T) = R(L, i, T) - C(L, i, T)$$

This is the most general formulation possible that still allows us to model differential effects on the revenue and cost sides.[7] Notice that to arrive at this simple statement of the problem, we have simplified away *all* of the details of the situation: it is the same if you're a bank, a person, a large retailer, etc.

If we have enough data with substantial variation, this problem can in principle be solved using supervised learning algorithms of the kind described in the Appendix. Why do we need variation? Since we are putting no structure into the problem—there are two unspecified functions—we need to let the data alone do the extra effort.

But most likely we're asking too much from our data, so we could proceed to simplify further. Suppose that for now we only care about price optimization, so we can fix the two other potential levers and ask ourselves how profits change with different interest rates:

$$\text{Profits}(i) = R(i) - C(i)$$

Let's continue with the data-driven solution and assume that the finance unit already has calculated profits, revenues, and costs at the loan level. We can then plot profits against interest rates and get a nice-looking profit function such as the one depicted in the left panel of Figure 5-6, where each dot corresponds to a hypothetical loan, all of the same size and maturity. I also plot a solid line corresponding to the relationship we uncovered with the data. Notice how lucky we were: the estimated profit function has a nice inverted-U shape, the kind we really want when solving a *maximization* problem. We can now proceed and solve the optimization problem by setting the interest rate to where we see that profits reach a maximum, which is around 10%.

---

7 We could just take the lefthand side of this equation and completely forget about revenues and costs.

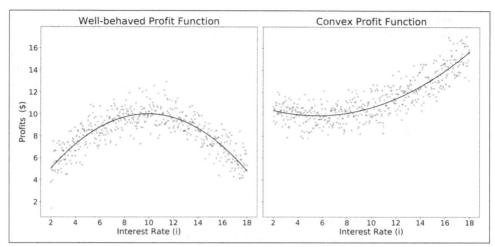

Figure 5-6. Two alternative scenarios when using the data-driven approach to estimate and optimize profits

However, since we did not impose any structure on the problem, we could very well have found the profit function in the right panel. The risk is that if we used numerical methods, we might mistakenly end up with a minimum close to 6% and make it our optimal interest rate. Note that the actual best rate might be even larger than 18%.

The lesson here is that oversimplifying can produce results that make no sense from the business point of view, and that can end up hurting the company. In this case we went too far, and we may now need to put some structure back into the problem. The best thing to do is to use our understanding of the situation to formulate a more precise problem.

Let's continue to simplify away the size of the loan and the maturity, and say that we are considering lending our customer $100 today that will be repaid tomorrow. If it's fully repaid, tomorrow we will get a revenue consisting of the principal and an interest payment. To do so we had to incur a cost to fund the loan, let's say at a lower, risk-free interest rate of i_s.[8] In this case, if the customer repays the loan, our profits will be the interest rate margin:

$$\text{Profits}(i) = 100(1 + i) - 100(1 + i_s) = 100(i - i_s)$$

---

8 Banks borrow at a low interest rate to gain on the margin obtained from lending at a higher rate. But even if your company doesn't *borrow*, you can think of this as the opportunity cost of putting those funds on any alternative investment opportunity.

Since this is a delinquency use case, we better allow for that possibility. Take the simplest case with only two possible outcomes: a borrower completely honors the loan, with probability p_p, or he defaults and, with some costly effort, we are able to recover a fraction k (at a cost $c(k)$ that is increasing in the fraction), with probability $p_s = 1 - p_p$. Notice that the funding cost was incurred with total certainty *before* the borrower decides to honor his debts or not.

Our expected profits are then:

$$\text{Profits}(i) = \underbrace{p_p 100(1 + i) + \left(1 - p_p\right)k100(1 + i)}_{\text{Expected Revenue}} - \underbrace{100\left(1 + i_s\right) - \left(1 - p_p\right)c(k)}_{\text{Expected Cost}}$$

This last equation bought for us a bit more realism, but surely not enough, since our expected profits are still *increasing* with the interest rates (our cost doesn't depend on how much we charge). We can approach this by either letting the probability be inversely related to the interest rate, or letting the cost $c(k, i)$ *increase* with the interest rate. Either of these assumptions will get us the trade-off we would expect to have: a higher interest rate still increases the expected revenues, but only at a higher cost.

Before moving on, notice that the extra structure allows us to estimate our expected revenues and costs, and therefore our profits: optimization is now feasible. If you wonder where our data fits in, we will use it to estimate customer probabilities of default or the collection cost function.

Stock Optimization

As was discussed in Chapter 3, first-order effects for inventory problems are given by the uncertain demand we face each period of time, production/ordering costs, transportation costs, and holding or storage costs.

Here I just want to show the value of simplification, and specifically, how different simplifying assumptions help us gain better intuition about the problem. Let's start by assuming away all uncertainty and costs. Since daily demand is known, we can just stock up whatever is needed for the next day (there are no transportation costs). But we could very well buy ahead for the next month, since there are no storage costs either.

Once we include these costs, we see that there are important trade-offs to consider. If transportation costs are high relative to storage costs, it will be optimal to overstock for several days. Alternatively, you might consider replenishing your inventory several times a day. You can see that that there are several important nuances in each simplifying assumption we make, and we gain some intuition about the first-order effects by relaxing one at a time.

Store Staffing

In Chapter 3 we mentioned that this problem is similar in spirit to that of finding an optimal inventory: by overstaffing we incur the cost of paying wages without the extra productivity, and understaffing may represent foregone sales or reduced customer satisfaction that, with some probability, will translate into higher customer churn and foregone earnings in the future.

But since this chapter is about understanding the value of simplification, consider the problem where we know with certainty how many customers will arrive each hour at each of our stores. We can then first solve the problem of finding the average waiting time for our customers.

Once we solve this highly simplified problem we can move one step ahead and relax the certainty assumption *but* impose strong distributional assumptions on the rates at which customers enter or exit the store that allow us solve the problem with uncertainty.

After this work has been done, we need to map waiting times to customer satisfaction, and again we may want to impose simplifying assumptions. The theory would be that satisfaction *increases* with lower waiting times, and we can start by assuming linearity, say. Alternatively, we can tie the probability of a customer churning to waiting times, which will directly impact our business objectives (our profits).

Here I just wanted to show how that we need theories to map actions to consequences that affect our business objectives, and the role that simplifying assumptions play in allowing us to solve a specific problem. I will show one possible solution in Chapter 7.

Key Takeaways

- *We impact our business objectives by making decisions*: as analytical thinkers, it is our job to find, test, and enrich the set of actions or levers we can pull to achieve our business objectives.

- *But our decisions can only indirectly change our performance*: there are intervening factors—human or technical—that we need to understand to be effective. These create consequences directly impacting the business.

- *The link between our actions and consequences is mediated by causation*: it is critical that we understand the causal forces from actions to consequences. To be effective we must also remember the possible traps when interpreting an effect causally.

- *As humans we have a powerful ability to create theories about how things work*: this skill is necessary to propose new levers that we can test through A/B testing.

- *The world is complex—we need to learn to simplify*: when creating theories that map actions to consequences, we will necessarily have to simplify away nuances of each problem. Ideally we should start by concentrating on first-order effects, but it's hard to know what is first-order from the outset. Domain knowledge is valuable, but also the ability to think out of the box to propose new theories.

Further Reading

If you're interested in exercising your analytical muscle by solving Fermi problems, you will find Lawrence Weinstein and John Adam's *Guesstimation: Solving the World's Problems on the Back of a Cocktail Napkin* (Princeton University Press) or Sanjoy Mahajan's *Street-Fighting Mathematics: The Art of Educated Guessing and Opportunistic Problem Solving* (MIT Press) useful. William Poundstone's *Are You Smart Enough to Work at Google?* (Little, Brown Spark) presents many examples of problems that are useful to accomplish the same goal, including a couple of Fermi problems.

To the best of my knowledge, there are no general books that teach how to simplify. Unfortunately this is a skill that you can only develop by studying models made by others and then trying to build your own. Scott Page's *The Model Thinker: What You Need to Know to Make Data Work for You* (Basic Books) can help you with the former, since he presents a variety of models used in the social sciences.

If you're interested in inventory theory or management, good bets are introductions to operations research or control theory (the latter for applications under uncertainty). Chapter 19 of Hillier et al., *Introduction to Operations Research* (McGraw-Hill Higher Education) presents a nice introduction to inventory theory with and without uncertainty.

## CHAPTER 6

Uncertainty

Benjamin Franklin is often quoted as saying, *"in this world nothing can be said to be certain, except death and taxes."* At the risk of stating the obvious, I would claim that *the only thing that is certain in life is uncertainty.* As we will see in this chapter, not only must we master the art of simplification, but we must also try our best to understand where uncertainty comes from, and how to make decisions when we're not sure about the results of our actions (Figure 6-1).

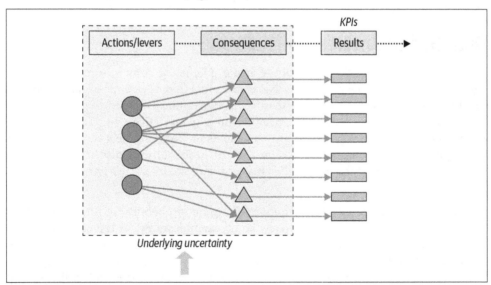

Figure 6-1. Understanding the underlying uncertainty

The main takeaway from this chapter is that to make decisions under uncertainty we will seek to maximize the mathematical expectation of our business result. We must therefore start by providing enough background on probability theory so that we feel comfortable with calculating expectations. I will also provide a primer into the theory of decision-making under uncertainty and finish by applying this toolkit to our use cases.

Where Does Uncertainty Come From?

Uncertainty reflects our ignorance about something. In the sciences, uncertainty or randomness are commonly associated with our lack of knowledge about the *causes* of some phenomenon. So where does it come from? We have already talked about this in Chapter 2, so let me just summarize what was mentioned there.

Many times, the source of our uncertainty reflects, literally, our *ignorance* about the causes or consequences of our actions or any phenomenon we observe. But it can also be a byproduct of our need to *simplify* away the complexities of the world that are not of first-order to the problem at hand. Other times we may be quite comfortable with our prediction about the direction or sign of an effect—think the Law of Demand— but we may be uncertain about the different magnitudes in a *heterogenous* population. Finally, uncertainty may arise from *complex* behavior and social interactions that may even obey simple deterministic rules.

This list is not supposed to be exhaustive, of course. But most uncertain events can be classified into one of these categories.

Quantifying Uncertainty

We are all equipped with the ability to make qualitative comparisons of uncertain events. For instance, given our experience and some other evidence we have, we may be able to say that *it is more likely* that it will rain tomorrow than that it won't. But how can we make this qualitative comparison a quantitative one? The calculus of probability was specifically designed to enable us to work and deal with uncertainty. I will give a quick and simplified summary of probability theory here, but please check the references at the end of the chapter if you are interested in more thorough treatments.

In our context, we are interested in quantifying the probability of uncertain consequences, given our actions. Let us view some examples now. In the context of customer attrition, we can make a retention offer to a customer, but she may or may not decide to churn. Here we are truly ignorant about two things: is the customer likely to churn even if we don't make an offer? Will our offer affect the probability of that event happening (and how)?

What about dating? Should I go out on that blind date tonight? In this case the action is going out or not. There are many uncertainties underlying this problem, but since we have one particular objective—say, meeting someone special that you can continue dating—we can simplify and consider only the probability that the date will be *successful* according to that objective. One thing should be clear from the outset: the probability of meeting someone significant is null *if* we decide to stay home.

To make things simple, consider only the case of binary outcomes where our actions have consequences that are either satisfactory or not.[1] There's no middle ground, and the two alternatives are mutually exclusive: one and only one must be true. Denote by $p(S)$ the probability of the consequence being *satisfactory* given our action. We will sometimes let the probability depend on our actions, and denote it by $p(S|a)$, read as the probability that the outcome will be successful, *given* that action a was taken. The following properties are typical of probability functions:

- $0 \le p(S) \le 1$

  This says that the probability is a number between zero and one—zero denoting the case where we know with absolute certainty that the event will *not* happen, and one the other extreme scenario.

- $p(S) + p(\text{not } S) = 1$

  Since in the binary outcome world the events are mutually exclusive, it must be that the two probabilities add up to one. Alternatively, we can express the probability of a failure (not a success) as one minus the probability of a success.

These two properties will suffice for us for now, but please note that the more general properties cover the cases where we have many possible outcomes and where some of them may not be mutually exclusive. The second property is known as an addition rule, and the calculus of probability also comes equipped with a *product* rule that allows us, among other things, to update probability estimates (Bayes' rule).

Bayes' Rule

Bayes' rule describes a way to update probabilities when new evidence comes in. Recall that $p(A|B)$ is the probability of A conditional on observing B. Bayes' theorem states that

$$p(A|B) = \frac{p(B|A)p(A)}{p(B)}$$

---

1 In most of our business examples a satisfactory outcome will be the one that we want to take place.

Here A, B are uncertain outcomes (events, in the language of formal probability), and $p(A)$ and $p(A|B)$ are the prior and posterior probabilities of event A happening, respectively.

It is commonly used to update probabilities in the light of new evidence: we start with some prior belief about an uncertain outcome we wish to predict (A) and observe evidence related to another event (B). Bayes' rule allows us to precisely update our beliefs by using the prior probability and the conditional probability.

Expected Values

Suppose you enter the following gamble: I will flip a coin and if it shows a head you *win* $10; otherwise you *lose* $10. How can you assess the value of this gamble should you decide to enter? Formally, we have defined a *random variable X* that takes two values depending on the uncertain outcome: 10 if it shows heads and –10 if it shows tails.

Notice that random variables are mathematical functions that map each element in the set of uncertain outcomes (tails or heads in our example) onto numbers (–10 and 10 in the example). If we had 5 possible outcomes (or 10, or infinite), each with an assigned probability, a random variable would assign each of the 5 (or 10, or infinite) outcomes a number. Also, since this is a mathematical function, each outcome must be assigned to one and only one number; however, several outcomes may be assigned the same number. In the extreme case—the case of a constant function—all outcomes are assigned the same number. Random variables allow us to translate random outcomes into numbers, which is quite handy since we know how to work with them.

In the case of our gamble, the uncertainty arises because we don't know if the coin will show heads or tails, so our prize is unknown. We can define the *expected value* of our gamble as the average of the prize, *weighted* by the probabilities:

$$E(\text{prize}) = p(tails) \times (-10) + p(heads) \times 10$$

If our coin is fair—so that each side has a 50% probability of showing—the expected value is $0.5 \times -10 + 0.5 \times 10 = 0$. Figure 6-2 shows the expected value for different probabilities of landing heads. Note that it varies *linearly* with the probabilities; a very nice property that will become handy in applications. When we are certain that it will land heads, we expect to get the full $10. At the other extreme, if the coin always shows tails, we have a sure loss of $10.

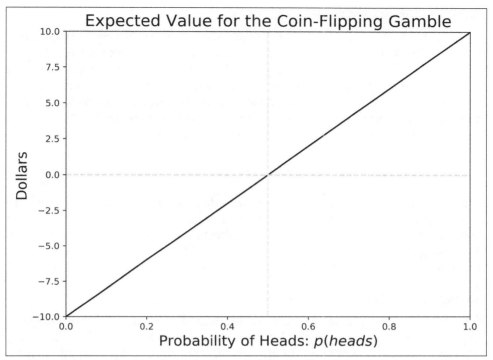

Figure 6-2. Expected values for different probabilities of heads for our coin-tossing example

Calculating Expected Values

In general, if our random variable takes N values $x_1, x_2, \cdots, x_N$, each one with probabilities $p_1, p_2, \cdots, p_N$ (note they must add up to one, as in the second property above), the expected value is defined as

$$E(x) = p_1 x_1 + p_2 x_2 + \cdots + p_N x_N$$

Simplifying a bit the exposition, in the case where we have an infinite number of outcomes (technically, *uncountably infinite*) we need to define the expectation with an integral.

Bidding for a highway construction contract

Let's explore another example. Suppose your company is considering participating in the new public bid to construct an interstate highway. The regulator has decided that any company that wants to participate must pay a fixed entry cost of $10K. The finance department has estimated that if the contract is awarded, the company's long-term incremental profits will be $1 million. After thorough consideration, the data scientists on the team estimated that the probability of winning the contract is 80%. Let's calculate the *expected profits* if we decide to participate:

$$E(\text{profits}) = 0.8 \times \$1M + 0.2 \times \$0 - \$10K = \$790K$$

Notice that you pay the entry cost *independently* of whether the company wins the contract or not, so it stays out of the expected value calculation. The only uncertain quantity is the value of the projected incremental profits with or without the contract. Note also that expected values preserve the units of the random variable under consideration. In this case, we expect to make $790K, should we enter the bid.

Interpreting expected values

How should we interpret the expected value? According to one school of thought—the *frequentist* school—we can think of an experiment as one that takes place *many* times under the same conditions. In the coin-flipping example, if you repeat the gamble an infinite number of times, the expected value is the (simple) arithmetic average of your long-run earnings: when you win (approximately 50% of the time over the long run) you add $10 to the long-run earnings, and when you lose you subtract $10. The simple average is obtained by dividing your total earnings by the number of times you entered the gamble.

Figure 6-3 shows an example of 1,000 realizations of this gamble, obtained with the code snippet that follows (Example 6-1). The left panel shows how total earnings accumulate over time: in this case, the player starts with a losing streak, and at least in the first 1,000 repetitions, they don't fully recover.[2] The right panel shows the average earnings, and you can see that they converge very quickly to the expected value of zero dollars.

---

2 This example is commonly known as the one-dimensional drunkard random walk, and it can be shown that our gambler will recover *infinitely many times*.

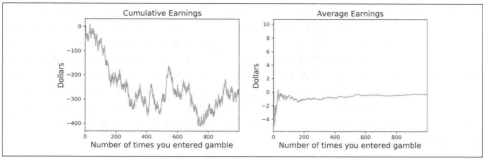

Figure 6-3. Repeating the coin-flipping gamble 1,000 times

Example 6-1. Average long-run earnings as expected values

```
import numpy as np
import pandas as pd
np.random.seed(1005)
# Initialize
N = 1000
total_earnings = pd.DataFrame(index= np.arange(N),columns = ['earnings'])
for i in range(N):
    # Draw from a uniform distribution
    draw = np.random.rand()
    # Heads if p>=0.5
    if draw>=0.5:
        curr_earn = 10
    else:
        curr_earn = -10
    total_earnings.earnings.loc[i] = curr_earn

# compute cumulative and average
total_earnings['cumulative_earnings'] = total_earnings.earnings.cumsum()
total_earnings['trials'] = np.arange(1,N+1)
total_earnings['avg_earnings'] = (total_earnings.cumulative_earnings /
                                  total_earnings.trials)
# ready to plot
fig, ax = plt.subplots(1,2, figsize=(10,4))
total_earnings.cumulative_earnings.plot(ax=ax[0])
ax[0].set_title('Cumulative Earnings', fontsize=14)
ax[1].set_title('Average Earnings', fontsize=14)
ax[0].set_ylabel('Dollars', fontsize=14)
ax[1].set_ylabel('Dollars', fontsize=14)
ax[0].set_xlabel('Number of times you entered gamble', fontsize=14)
ax[1].set_xlabel('Number of times you entered gamble', fontsize=14)
total_earnings.avg_earnings.plot(ax=ax[1])
plt.tight_layout()
```

As simple as the frequentist interpretation is—so called because you can imagine probabilities denoting the frequency of occurrences if you repeat the gamble infinitely many times—it may be problematic. What does it mean to repeat a gamble many times, if we know that it will happen only once? Even worse, think about the case of the public bid. How can we even imagine entering many times, in the *exact* conditions of the current bid? Proponents of the *Bayesian* school of thought argue that there are many cases where the infinite-repetition interpretation is not natural; they claim that in these cases, each individual may have a different probability assessment that need not coincide with a long-run frequency of occurrences. In this case, subjective probabilities quantify the degree of uncertainty that any of us may have, and may, therefore, not coincide at all.

Bayesian and Frequentist Schools of Thought

The two schools of thought mentioned in this chapter are the classical or frequentist and the Bayesian schools. The 20th century witnessed very strong debates between the two with respect to their interpretation of probabilities. Broadly speaking, the classical school thinks of probabilities as long-run frequencies of occurrences, and Bayesians believe that a probability is a subjective assessment of uncertainty. It is called *Bayesian* because, in their view, we all start with prior subjective beliefs that are updated through Bayes' rule once we get new evidence.

Making Decisions Without Uncertainty

Making decisions under uncertainty is hard because we must first master the calculus of probabilities. A useful trick is to always start simple, assuming away any underlying uncertainty: if we knew everything that is relevant to our problem, how would we choose? This helps us clarify many things, starting with whether we actually have levers that can accomplish our objectives.

Consider first the decision in Figure 6-4. We want to get the largest revenues possible, and we are considering two alternative actions: we can either make a price discount, in which case our revenues will be $155K, or we can fuel our online marketing campaigns, generating $131K. All of these quantities are known since we are assuming away all uncertainty.

What should we do? In this case things are straightforward and we decide to go with the price discount. This follows from the fact that we have chosen revenues to be our objective, and that we seek to generate the maximum revenues possible.

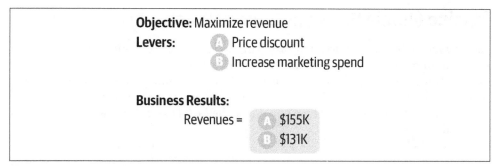

Figure 6-4. Choosing a lever under certainty

Decisions without uncertainty are *relatively* easy to solve, but this is most likely an artifact of our (hopefully) conscious choice to simplify. To see this, consider the two cases shown in Figure 6-5. We have two levers (A and B) and two objectives: revenues and customer satisfaction. The ideal world is the one in the left panel, where both objectives are aligned and we choose action A since it is superior to B for the two metrics under consideration. But many times the objectives look more like those in the right panel, and we need to trade-off one objective for another.

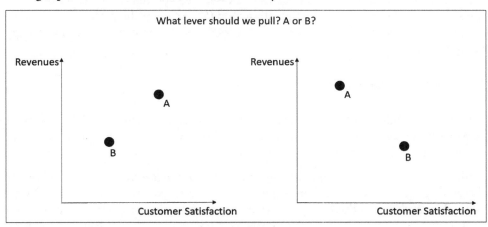

Figure 6-5. Decisions with multiple objectives

How should we attempt to solve this problem, even in the case where we have complete certainty about the mappings from our actions to the consequences? One possibility is to try to measure everything with the same standard, say, dollars. Can we translate customer satisfaction into dollars? There's another uncertain mapping that we need to quantify, but if we succeed then we are comparing apples to apples, and we get one simple dimension to optimize.

Making Simple Decisions Under Uncertainty

Let us modify the simple decision problem in the last section, and now assume that our customers may or may not accept our price discount (Figure 6-6). They accept with 80% probability and reject our offer with the remaining 20% probability. In the former case, we make $155K in additional revenues, and we make nothing if they reject our offer. Our second lever remains as before. How should we approach this decision?

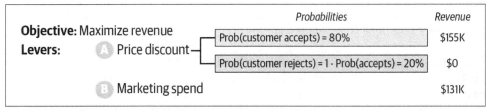

Figure 6-6. Same decision, but now under uncertainty

Let's compute the expected revenues of our two levers, starting with the marketing campaign since it has no uncertainty. In this case, the expected revenues are the revenues observed *if* we decide to pursue this action ($131K). For the price discount campaign, the expected value is revenues of $124K dollars, as can be readily seen by computing the expectation:

$$E(\text{revenues}) = 0.8 \times \$155K + 0.2 \times \$0 = \$124K$$

Which offer would you make? Comparing the uncertain discount ($124K in expectation) with the certain marketing campaign ($131K), we see that the latter has larger revenues and decide to move forward with it.

At this point you may be tempted to say that *under any circumstance* you would prefer the certain action (marketing campaign) since it's better to have this "mediocre" outcome than nothing. In this case you may be acting as a *risk-averse* person—in a somewhat extreme version—and, consequently, you prefer the safe bet. But let me try to challenge this position by asking the following question: given these probabilities (80–20), is there any set of rewards that will make you accept the bet?

Suppose that our data scientists reestimate the revenues from the price discount campaign at $164K. You can compute the expectation again (0.8 × $164K = $131.20K), compare it to the certain outcome, and see that now, under this expected value criterion, you should choose the price discount lever. Figure 6-7 shows how the expected value varies with different possible revenue estimations for this campaign. Are you still unconvinced?

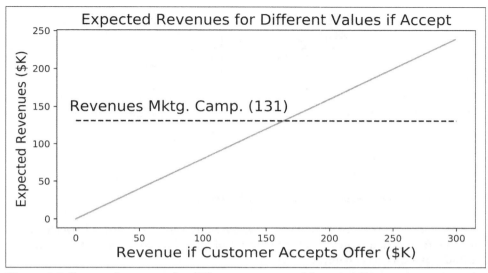

Figure 6-7. Expected revenues for different outcomes of the price discount campaign

What if our data scientists now estimate that the revenues from the discount campaign are $200K (expected value = $160K)? Or $500K ($400K in expectation) or $1 million ($800K in expectation)? I doubt that you are still reluctant to enter the bet, but we will talk more about this later in the chapter.

Expected values allow us to compare uncertain outcomes, so this will be our criterion when facing decisions under uncertainty. Let us now delve deeper into the difficulties when making decisions under uncertainty.

Risk Aversion

You may have heard about the concept of *risk aversion*. Informally, it captures the idea that we may be reluctant to enter a bet when there is a certain alternative. To quantify our degree of risk aversion, we start with a utility function that allows us to measure how we value different outcomes. $u(x)$ denotes the utility we derive from the outcome x. Utility functions convert the units of this possibly uncertain object (money, a car, a potential spouse, a potential new employer) into units of value (sometimes called utils).

Note that in all of our examples we've encountered so far we've assumed that the utility for money is *linear* ($u(x) = x$)—so that we value each dollar at face value; this is the case of *risk neutrality* and is captured by linear utility functions (Figure 6-8).

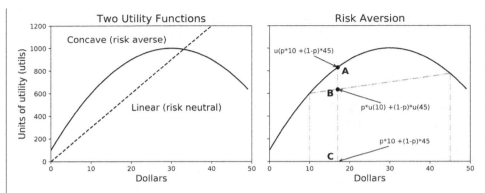

Figure 6-8. Risk-averse and risk-neutral utility functions

Concave utility functions exhibit *risk aversion*. Intuitively, if you are risk averse then you prefer a certain outcome over the result of a gamble since you'd really want to avoid making a decision under uncertainty.

You can see this in the right panel of the figure. With concave utility functions it is always the case that we prefer the certain outcome (point **C**) $px + (1 - p)y$ to the bet where we get $u(x)$ with probability p and $u(y)$ with the remaining probability. In the figure, note that point **A** is higher than point **B**, so the utility from the certain outcome (**C**) is higher than the expected utility of the gamble.

As will be shown later, in certain business applications we must be careful about modeling risk-averse utility functions.

Decisions Under Uncertainty

The main difficulty with uncertainty is that we can't be sure about the outcome of our decision at the time we choose our actions. As it is commonly said, *uncertainty unravels once we've made a decision*, at which point we might regret our choice if the realization was not satisfactory.

Figure 6-9 shows the main difficulties and a solution when we make decisions under uncertainty. Starting from the top-left, Panel 1 shows our choice and how our levers A, B map to the uncertain consequences (accept/reject each offer). They accept lever A with probability p_A and lever B with probability p_B. This is the underlying uncertainty in this hypothetical scenario.

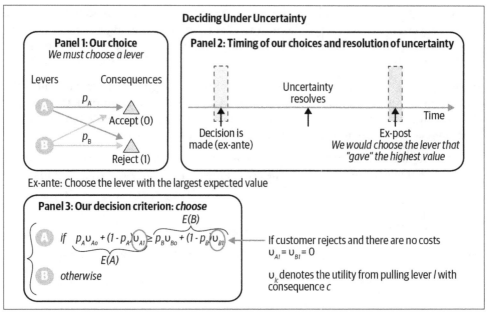

Figure 6-9. Anatomy of a decision under uncertainty

Panel 2 shows the main problem we face: we must make the decision *before* the uncertainty is resolved, commonly referred as the *ex-ante* stage. Ideally, if we had been able to see the actual outcome after uncertainty is resolved, we could have made the *ex-post* optimal decision, but this is not how decisions under uncertainty work. We must deal with the fact that decisions are made *before* uncertainty is resolved.

Panel 3 shows how we make choices under uncertainty: we evaluate each of the alternatives (our levers) by computing their expected values and choose the one that maximizes the expectation. It must be clear that this does not guarantee ex-post optimality: we may choose the option that maximizes the expected value, but still end up with a suboptimal outcome once uncertainty is resolved.

Why can this happen? It could very well be that we didn't have good estimates of our probabilities, either because of lack of good data, because we just didn't make the best use of our machine learning toolkit, or because we didn't spend enough time understanding the sources of our underlying uncertainty. But *even if we did*, sometimes we really just have bad luck.

Is This the Best We Can Do?

The expected value methodology has become the standard when making decisions under uncertainty. But is this the best we can do ex-ante, or before uncertainty is resolved? Let's explore some alternatives.

Suppose that instead of pulling the lever that maximizes the expected value, we neglect uncertainty and always choose the alternative with the highest payoff. If you only encounter this decision once, sometimes you'll regret it and some times you won't: it really depends on the values of the probabilities and the exact realization.

Since repeating the decision once wasn't conclusive, imagine that we encounter the exact same bet many times as a frequentist would hold. If you compute the *cumulative* earnings for both methods, you would then be happy to conclude that maximizing the expected utility is unconditionally better in the long run.

Figure 6-10 shows the cumulative earnings from these two decision criteria if we faced the same decision 100 times. The code shown in Example 6-2 was used to create the plot:

Example 6-2. Maximizing expected values is (second-best) optimal in the long run

```
def get_exante_earnings(accepts_a, accepts_b, exante_choice, clv_a, clv_b):
    '''
    Our earnings depend on customers' acceptance of each offer and
    Expected Utility
    1. If E(A)>E(B): we choose to offer alternative A
       If customer accepts A we make CLV_A otherwise we make 0
    2. If E(A)<E(B): we offer B
       If customer accepts B we make CLV_B otherwise 0
    '''
    earn_ea = 0
    if accepts_a == True and exante_choice=='a':
        earn_ea = clv_a
    elif accepts_b == True and exante_choice=='b':
        earn_ea = clv_b
    return earn_ea

def get_expost_earnings(accepts_a, accepts_b, clv_a, clv_b):
    '''
    Ex-post is first-best: we choose as if there was no uncertainty
    1. If customer accepts A and not B -> offer A
    2. If customer accepts B and not A -> offer B
    3. If customer accepts both -> offer the best for us
    '''
    earn_ep = 0
    if accepts_a == True and accepts_b ==False :
        earn_ep = clv_a
    elif accepts_a == False and accepts_b ==True :
        earn_ep = clv_b
    elif accepts_a == True and accepts_b ==True :
        earn_ep = np.max(np.array([clv_a, clv_b]))
    return earn_ep

def get_maxvalue_earnings(accepts_a, accepts_b, clv_a, clv_b):
    '''
```

```
    Rule: if CLV_A>CLV_B -> offer A (otherwise B)
    Earnings: we only make corresponding CLV if it aggrees with acceptance
    '''
    earn_mv = 0
    if clv_a>=clv_b and accepts_a ==True:
        earn_mv = clv_a
    elif clv_a<=clv_b and accepts_b == True:
        earn_mv = clv_b
    return earn_mv


np.random.seed(7590)
# Customer lifetime values if accepts (otherwise 0)
clv_a = 10
clv_b = 11
# acceptance probability
prob_a = 0.6
prob_b = 0.5
# expected values and optimal choice under expected utility
evalue_a = prob_a*clv_a + (1-prob_a)*0
evalue_b = prob_b*clv_b + (1-prob_b)*0
if evalue_a> evalue_b:
    exante_choice = 'a'
else:
    exante_choice = 'b'
# simulate T choices: earnings initialized to 0
T = 100
total_earnings = pd.DataFrame(index=np.arange(T),
                        columns=['exante','expost','max_prob','max_value'])
for t in range(T):
    # Simulate uncertain choices by our customers
    accepts_a = np.random.rand() <= prob_a
    accepts_b = np.random.rand() <= prob_b
    # Ex-ante Optimal:
    total_earnings.exante.loc[t] = get_exante_earnings(accepts_a, accepts_b,
                    exante_choice, clv_a, clv_b)
    # Ex-post optimal:
    total_earnings.expost.loc[t] = get_expost_earnings(accepts_a, accepts_b,
                    clv_a, clv_b)
    # Always choose max_value
    total_earnings.max_value.loc[t] = get_maxvalue_earnings(accepts_a,
                    accepts_b, clv_a, clv_b)


# ready to plot
fig, ax = plt.subplots(1,2, figsize=(12,4))
total_earnings.expost.cumsum().plot(ax=ax[0],color='k', ls='-',lw=5)
total_earnings.exante.cumsum().plot(ax=ax[0],color='k', ls='--')
ax[0].set_title('Cumulative Realized Earnings', fontsize=16)
total_earnings.max_value.cumsum().plot(ax=ax[0],color='k', ls='dotted')
df_relative_earnings = pd.DataFrame(total_earnings.max_value.cumsum() /
                    total_earnings.exante.cumsum(), columns=['ratio'])
df_relative_earnings.ratio.plot(ax=ax[1],fontsize=16, legend=None, color='k')
ax[1].plot([0,100],[1,1], ls='--', alpha=0.5, color='0.15')
```

```
ax[1].set_title('Ratio of Max Value to Ex-Ante', fontsize=16)
ax[1].set_xlabel('Number of times you make the same decision', fontsize=12)
ax[0].set_xlabel('Number of times you make the same decision', fontsize=12)
ax[0].set_ylabel('Dollars',fontsize=12)
ax[1].set_ylabel('Dollars',fontsize=12)
plt.tight_layout()
```

The ex-post decision is sometimes called the *first-best*. This is because we effectively choose assuming away any uncertainty, as if we had a magic ball and know what the customer will accept. As such, this is always a good benchmark to compare any other decision criterion. I also show the simulated earnings for two additional criteria: maximizing the expected value (labeled *Ex-ante*) and always choosing the alternative with the highest certain payoff (labeled *Max-Value*, option B in our example).

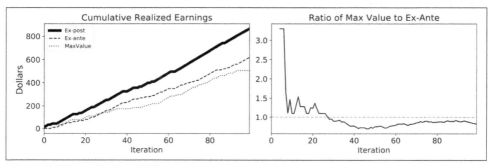

Figure 6-10. Evaluating different decision criteria under uncertainty

The left plot shows that the first-best is indeed the best possible outcome. In this simulation, choice B had a few good initial realizations, making it superior to the expected value calculation ("Ex-ante") in the short term. But in the longer-term this behavior reverts, and had we been using the expected-value calculation, we could've made more revenues. The answer is that with this set of realizations, *we would've regretted not using the expected value criterion*.

The right plot shows exactly the fraction of extra earnings we could've made. At the beginning the few good initial realizations for choice B (with higher CLV) made the MaxValue criterion superior to the use of expected values. Since the frequency of occurrences converge to the probabilities in the long run, we start seeing that the ex-ante criterion starts dominating.

Note that a similar argument can be made if your choice was based only on probabilities—say, always choose the lever with a higher probability of being accepted.

But This Is a Frequentist Argument

The previous example shows that maximizing the expected value is better than the two alternatives we have proposed: neglect uncertainty and choose the lever that gives the highest revenues if accepted, and neglect revenues and choose the lever with highest probability of acceptance. In both cases, we would have regretted, *in the long run*, using these decision criteria.

But I argued previously that this frequentist interpretation may not be suitable in many realistic business scenarios: the whole idea of simulating many times the same decision *under the exact same conditions, only with different realizations of uncertainty* is problematic.

Decision theorists have battled with this dilemma for decades, and the answer has been the so-called axiomatic approach. Under this solution we need not assume a frequentist rationale, but rather pose some behavioral axioms that when satisfied guarantee that we act *as if we maximized the expected value*. I will mention some references to this approach in the Further Reading section, but let me just finish by saying that it doesn't follow from the axiomatic approach that maximizing expected values is the absolute best we can do. The axiomatic approach guarantees, however, that maximizing the expected value is the rational thing to do.

Normative and Descriptive Theories of Decision-Making

The expected value criterion can be thought of as a theory of decision-making under uncertainty. When presented with a decision that involves uncertain outcomes, how *do* I decide? And how *should* I decide? These are two very different questions that we have already encountered: the former *describes* what is done and the latter provides a *prescription* or recommendation on what is the best course of action.

Since in our day-to-day choices most of us do not make expected value calculations, this necessarily implies that the theory of expected utility—as it's called—isn't an accurate description of how we make decisions. But is it a good *normative* theory, that is, if we were able to do such calculations, would we be better off? As discussed in the previous section, under the frequentist interpretation the answer is positive. *We will make better decisions for our companies if we adopt the expected utility criterion*. This is the reason why we include it in our analytical toolkit.

Some Paradoxes in Decision-Making Under Uncertainty

Suppose you are presented with the following gamble: you can win a prize of $1 million with a probability of 0.001 and nothing otherwise. What is the maximum you should be willing to pay for this lottery ticket? As you might've expected (no pun intended), it's the expected value of the gamble.

Let's compute the *expected profits* for any given price we pay to participate (y):

$$E(\text{profits}) = 0.001 \times \$1M - y \geq 0$$

The last inequality just states that for us to participate it should be better that we don't lose in expectation. It follows that the maximum price we should pay can be found when we break even: $y^M = \$1,000$.

To make this example more realistic, let's think about the Mega Millions lottery in the US, and take the probability of our lottery ticket being one of the lucky ones to be 1 in 302,575,350.[3] At the time of writing, the cost of each ticket is $2. At this price, the minimum jackpot for which we should be willing to play is $605M.[4]

Many people play the lottery even when it's not optimal according to the expected value criterion, so again, as a descriptive theory of decision-making it seems that it doesn't do a great job.

Now think of the following lottery: to participate you must pay all of your savings (say $100). With probability of $1E-6$ you win 1,000,001 times your savings. With the remaining probability you don't win anything (and thus you lose everything).[5]

$$E(\text{prize}) = 0.000001 \times (1000001 \times \$100) = \$100.0001$$

The probabilities and prizes are chosen here so that, according to the expected utility criterion, it is *always better* to enter the gamble, independently of your savings. But would you enter? I know I wouldn't.

This leads to the most famous paradox in decision theory concerning uncertainty.

The St. Petersburg Paradox

Suppose you are offered the following gamble: I will throw a fair coin—so that heads and tails are equally likely—and give you 2^n dollars if the first head appears on the n-th toss. Since this is a fair coin, the probability of a head appearing on the first toss is $1/2$, on the second $(1/2)^2$—with probability $1/2$ the first toss shows a tail, and with the same probability, it shows the first head—and so the probability of the first head appearing on the n-th toss is $(1/2)^n$.[6] Let's compute the expected value for the prize:

$$E(\text{prize}) = \frac{1}{2}2 + \frac{1}{4}4 + \frac{1}{8}8 + \cdots = 1 + 1 + 1 + \cdots$$

---

3 In October 2018, these were the odds of winning the Mega Millions lottery in the US. See this article on CNBC's website (*https://oreil.ly/VJh4W*).

4 It's even higher, as there could be more than one winner and you would have to split the jackpot.

5 This example was taken from E.T. Jaynes's book. See the Further Reading section at the end of the chapter.

6 Just to be sure you're following, the key thing to notice is that each toss is *independent* of the others, so that the probability of a sequence such as TTTTH arising after five tosses is the product of the probabilities of each occurrence.

As you can tell, the probabilities and prizes are chosen so that the expected prize grows without bound. The paradox arises because no one would be willing to pay such an amount to enter this bet—usually called the *fair price* for the lottery.

In the 18th century, mathematician Daniel Bernoulli proposed a solution: we shouldn't value each prize at face value, rather, we should be using a utility function that displays the diminishing value that each extra dollar represents to us. The solution he proposed was to value each prize with the natural logarithm (a nice, concave function, thereby displaying diminishing marginal utility, and as we've already learned, risk aversion).

Figure 6-11 shows the expected values for the two alternatives. As shown previously, the expected utility for the linear case grows one-to-one with the number of tosses. On the other hand, the expected utility for the log case converges nicely to under 1.4 utils (the unit of measurement for utility), which is reached after around 15 realizations. It follows that we should pay to participate in at most 15 rounds of the bet. The incremental value after that is zero for us.

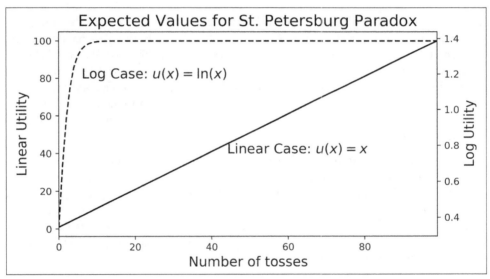

Figure 6-11. Expected values using a log and linear utility function

The paradox is important for us, because up to now we have been valuing our utility at face-value with the dollars we make. We have even found the maximum value—the fair value—using expected value calculations. The paradox reminds us that we should be careful when we use risk-neutral utility functions, as we may end up broke.

Risk Aversion

As the St. Petersburg paradox reminds us, sometimes it is important to model our choices with concave utility functions, as they not only display diminishing marginal utility but they also display risk-averse preferences. We briefly mentioned risk aversion previously, but it wasn't clear why it matters in business applications, so let's revisit the example in Figure 6-6 but change the probabilities (Figure 6-12).

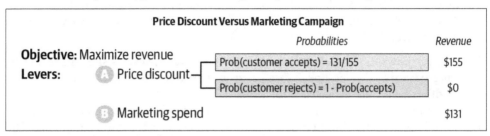

Figure 6-12. Revisiting the discount versus marketing campaign example

You can now check that the expected value of the risky decision equals the sure revenue from the marketing campaign. Strictly speaking, since both levers have the same mathematical expectation, we should be *indifferent* between the two. We could choose randomly between them, always the first, always the last, or use some other tie-breaking criterion that is extraneous to the expected value calculation. I don't know about you, but I don't feel comfortable with this recommendation: in my case I'd always prefer the certain outcome, so it would be amazing that this analytical machinery actually recommends that course of action.

You may recall that up to now we have used linear (risk-neutral) utility functions that map each dollar to one unit of utility ($u(x) = x$). But the previous discussion suggests that I'm not really risk neutral; as a matter of fact I'd avoid entering the gamble and get the certain revenue. So here lies the answer to our puzzle: we should replace our risk-neutral utility function with a concave one that better captures our preferences for risk. For the sake of the presentation let's use Daniel Bernoulli's proposed solution that maps dollars to log dollars.

For the uncertain price discount we get:

$$E(\text{Revenue}|\text{Discount}) = \frac{131}{155} \times \log(155) = 4.3$$

And we already know how to compute the expected revenue for the marketing campaign:

$$E(\text{Revenue}|\text{Mkt.Campaign}) = 1 \times \log(131) = 4.9$$

It now follows that we should go ahead with the marketing campaign. If you're wondering if this will work for any other concave utility function, the answer is *yes* as it follows nicely from the definition of concavity.

The good news, then, is that we can keep using the expected utility criterion to solve business problems. The bad news is that we have added another level of complexity and we may now need to choose a risk-averse utility function for our applications: we get crisper decisions but only at the cost of losing linearity. My recommendation is to start simple and assume risk neutrality (the linear world is just easier to deal with to begin with). Once you understand this simpler problem, you can try your best to understand the preferences for risk of your stakeholder, and if necessary do some calibration. If you're interested, some utility functions commonly used to model risk aversion are listed in Equations 6-1 through 6-3. The free parameters can be used for calibration.

Equation 6-1. Logarithmic utility function (no free parameters)

$$u(x) = \ln(x)$$

Equation 6-2. Polynomial utility function

$$u(x) = x^a \text{ for } a \in (0, 1)$$

Equation 6-3. Exponential utility function: constant absolute risk aversion (CARA)

$$u(x) = 1 - e^{-ax} \text{ for } a > 0$$

We can combine a normalized version of the polynomial and the log utility functions to create a *constant relative risk aversion* function commonly used by economists. Coefficients of risk aversion depend on the relative curvature of the utility function. For instance, in the exponential case, a is the coefficient of absolute risk aversion. In the CRRA case shown in Equation 6-4, ρ is the coefficient of relative risk aversion. Please see the references listed in the Further Reading section if you're interested in understanding the differences between the two.

Equation 6-4. Constant relative risk aversion (CRRA)

$$u(x) = \frac{x^{1-\rho}}{1-\rho} \text{ for } \rho \neq 1 \text{ and } \ln(x) \text{ for } \rho = 1$$

Figure 6-13 shows how some of these alternatives look for different parametrizations. As you would have expected, different alternatives and parametrizations change the curvature of the utility functions, thereby providing systematic ways to model our preferences for risk.

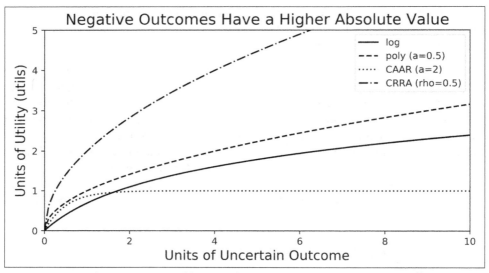

Figure 6-13. *Different ways to calibrate risk aversion*

Putting it All into Practice

By now I hope I've convinced you that the expected value criterion is a powerful, yet simple method to make decisions under uncertainty. I also hope I was able to convey some of the difficulties one may encounter. The simplicity comes from the fact that mathematical expectations are linear in the probabilities, and if you assume risk neutrality, also in the values for the uncertain outcome.

Let's summarize the approach now:

Decision-Making Under Uncertainty

If we have two levers (A, B) with uncertain consequences, we choose the one that maximizes the expected value of the metric under consideration:

- Choose A when $E(x \mid A) \geq E(x \mid B)$.
- Otherwise, choose B.

The same principle generalizes to cases with more than two levers.

What role does AI play in this calculation? There are two approaches that can be followed depending on the type of problem you're facing:

- You can directly estimate the expected values using the ML toolkit (supervised regression models).
- You can estimate the probabilities in the expected value calculation (supervised classification models).

I haven't defined these models in the body of the text, but please consult the Appendix to get more information. I will start by describing the latter approach.

Estimating the Probabilities

Expected values depend on probabilities and rewards for uncertain outcomes. We will first look at methods to estimate the probabilities.

Estimating unconditional probabilities

Going back to the frequentist interpretation of probabilities, we may want to start by estimating unconditional frequencies of occurrences. We have already discussed some of the problems with this method, but if we have historical data, it may be the way to start thanks to its simplicity. As a matter of fact, even if you do not plan to use this method, it is always good to start by plotting some frequencies of occurrences to gain some understanding of your data.

Figure 6-14 shows how such an approach may look for hypothetical historical data on conversion rates for retention campaigns. Here we see that, historically, 20% of our customers have accepted our retention campaigns. We can therefore use this as our base estimate to compute the expected value if we wanted to choose this lever or any other we may be analyzing.

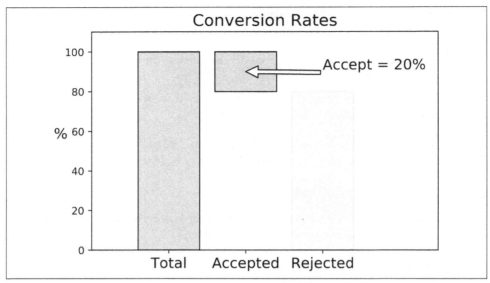

Figure 6-14. Historical conversion rates for retention campaigns

The upside is its simplicity: if we have the data (that may be a big "if") computing frequencies is almost immediate. There are several drawbacks, however. Assuming we have data, since we are pooling the information from possibly different campaigns with different samples of customers, we have no way to control for these differences. For instance, is the holiday campaign the same as our recurrent retention offers? Are the customers the same? The answers to these questions are most likely in the negative. Since we are not controlling for these factors, these are generally called *unconditional* probability estimates.

One may start filtering out some campaigns, or slicing and dicing your data to get different estimates that may look similar to the ones in the figure, and of course, that will help your analysis. But other methods may prove more powerful if you want to control for observables in a systematic way.

Estimating conditional probabilities

As described in the Appendix, the two main classes of methods in supervised machine learning are regression and classification. Classification is commonly used to predict discrete categories like the ones used in the previous examples (customer accepts or rejects an offer, churns or remains loyal, etc.). Leaving the details to the Appendix, to get these predictions we estimate conditional probabilities of these categories being true; these *conditional probability estimates* allow us to move away from broad generalizations and get closer to the realm of customization.

For instance, instead of using the aggregate churn rate, we can condition on the tenure of each customer (as it is quite common that customers that have been loyal to our company are less likely to switch companies, and vice versa). One can then estimate a classification model that depends on your customers' tenure and other controls you consider relevant (Equation 6-5).

Equation 6-5. Conditioning on the tenure of your customers

$p(\text{churn}|\text{tenure in months}) = f(months)$

If you allow for enough nonlinearity, you may end up with something like the pattern shown in Figure 6-15. In this hypothetical scenario, customers that have been loyal for 18 months are the least likely to switch companies, but note that around 15% still leave.

Also notice that I'm not reporting any estimates of how uncertain we are about this hypothetically fitted function. It is a good practice to do so in your own work. I'll have something to say about this later.

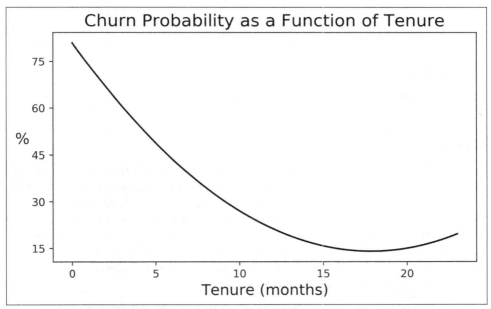

Figure 6-15. Hypothetical estimated conditional probability of churning

The important thing to remember is that classification models allow us to make conditional probability assessments. By conditioning on some observables we not only get a clearer picture about the heterogeneity of our sample of customers, but it also allows us to go one step toward customization of optimal decisions. We are no longer pooling every customer in our sample into the same bag, but we are starting to customize our estimates, and thereby our decisions.

A/B testing

In Chapter 2 I described some of the risks of using observational data, and in particular, I mentioned the fact that when customers self-select (or are selected by us) our estimates may display significant bias. The experimental method (A/B testing) allows us to eliminate the selection effect and provide better estimates.

Putting away some nuances that I will briefly touch upon later, operationalizing A/B testing to get probability estimates is straightforward. Once our test is over you can either use the frequency analysis or the conditional analysis already described.

To summarize the pros and cons of testing, on the upside we already know that it allows us to estimate causal effects by eliminating the selection bias that is pervasive in observational data, and it is relatively easy to implement once you know how to choose sample sizes. On the downside, it is worth mentioning that A/B testing only allows us to estimate average effects for a sample of our customers (that we intend to extrapolate to the whole universe), so how to obtain personalized estimates for individual customers is not immediately clear. Also, there are times where you simply can't do A/B testing or it's forbiddingly costly (e.g., a customer churn experiment).

Bandit problems

Bandit problems are a class of *sequential decision problems* where we must make a choice that repeats over time, and as time goes on we are learning the workings of the underlying uncertainty, either by improving our probability estimates or the expected values themselves.

I won't go into the details, but I just want to mention the main trade-off we face with these sequential problems. The idea is simple: each time you make a choice to pull a lever, the outcome reveals some information about the underlying uncertainty, say the probabilities that our customers accept. Suppose you pull the price discount lever and find that your customers' acceptance rate is 80%, making the expected revenue for this lever higher than a competing one. Since you have only made the decision once, there's still considerable uncertainty about the estimated probabilities (put differently, maybe you were just very lucky on that first trial).

You now face a decision: you can keep pulling or exploiting the discount lever, or you can explore the untouched lever. It is not uncommon that our business stakeholders will pressure us to choose the former (first result was very good), but it may be better to experiment with the other one for a while and get better probability estimates. This is the famous *explore-exploit* trade-off pervasive in sequential problems of decision-making under uncertainty.

This is a fascinating topic that deserves a book of its own, so I'll stop now and provide references at the end of the chapter.

Estimating Expected Values

The expected utility hypothesis says that under uncertainty we are better off choosing whichever lever maximizes the mathematical expectation of the object under analysis. We may use our machine learning toolkit and estimate the probabilities as described in the previous section, or we can directly estimate the expected values. Probabilities are instrumental to computing expected values, but depending on the problem, we may just skip this step and attempt to estimate the expectations.

If the metric we wish to predict is continuous—like revenues, profits, lifetime values, or the like, we can then use *regression* algorithms. I'll leave the more technical material for the Appendix, but just let me mention that statistically motivated regression algorithms usually allow us to estimate the conditional mathematical expectation of the object of interest: if the variable is continuous, we may end up with the necessary estimates to make our decision, and if it's categorical, the outcome will be an estimate of the conditional probability.

Frequentist and Bayesian Methods

We have already encountered some of the differences between the Bayesian and frequentist (or classical) schools of thought, starting with the alternative interpretations of probabilities and expected values. One way to put it is that the Bayesian interpretation has a bottom-up feeling while frequentists approach the problem from a top-down perspective.

In classical statistics, probabilities are an objective truth in nature, and we can therefore imagine repeating an experiment many times. In this case, the relative frequencies of occurrences *converge* to the corresponding probabilities, like in the coin-flipping example. It is not a coincidence that limit theorems—what happens when we repeat an experiment an infinite number of times—are of fundamental value for this school of thought (think of weak or strong laws of large numbers). In a sense, we *uncover* these probabilities by repeating an experiment under the exact same conditions.

Bayesians, on the other hand, build their theory from the ground up, starting with each individual assessment of the likelihood of something taking place. As such, they seek conditions (axioms) such that this ignorance can be quantified with the standard calculus of probability (including Bayes' theorem, from which the label is derived). Probabilities here are subjective in the sense that two people may have different assessments on the likelihood of an event, and need not agree with any long-run frequencies.

I have tried my best to give a practical introduction to the topic, with the hope that the interested reader can go to the references and fill in the many details left out, and as such, I would do no justice to this topic if I try to go deeper in just a few paragraphs. I do want to say that we need to think hard about where uncertainty comes from and how to model it.

Take the example of how the churn probability varies with the tenure of our customers (Figure 6-15). The pattern in the data is hypothetical, but it is not unlike what you would find in real-life data from your business. But is there really evidence that our customers behave like this?

In most applications we report confidence intervals when describing this type of evidence, and at a minimum we should do so. But let me just point out that there are many difficulties when you try to interpret these intervals from a probabilistic perspective (as well as with p-values). Again, I'll do some hand-waving and point to the references at the end of the chapter.

Revisiting Our Use Cases

It's time to start analyzing each of our selected use cases. In what follows keep in mind that I have simplified away many uncertainties that are not of interest or of first-order for each specific problem.

Customer Churn

Consider first the case of making a single retention offer—that is, we may neglect for now the case of competing offers since, from the point of view of the intrinsic uncertainty, all may be analyzed in the same way. The underlying uncertainty in this scenario is whether our customers churn or not, possibly depending on whether we made the offer or not, as seen in Figure 6-16.

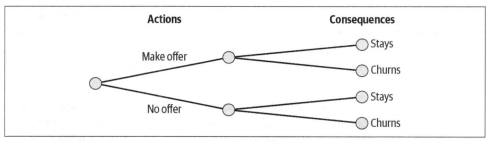

Figure 6-16. Underlying uncertainty: customer churn

Let's delve a little bit deeper into the sources of uncertainty here. To start, we don't know if a customer is likely to leave. As mentioned in Chapter 4, customers usually like quality, price (offer), and customer experience, and are generally willing to trade these off, at least to a certain degree.

So our first source of uncertainty is their satisfaction relative to their expectations on each of these dimensions. We also don't know how they rank them—that is, how sensitive they are to changes in each one—and to what extent they're willing to substitute one for the other.

To give you an idea of how we can model uncertainty here, let's simplify and consider first only the case of one dimension, say, quality. It sounds reasonable to imagine that customer satisfaction increases with quality: mathematically, you can think of a utility or satisfaction function that depends on their perception of quality, $u(q)$, and this function is increasing. We can also include some of the findings in the behavioral economics literature (see Chapter 4), and imagine that customers judge perceived quality relative to their expectations or a reference point, say q_0, so that $u(q - q_0)$.

We now have a theory of how satisfaction varies with quality, and we may proceed to link this to their decision to switch companies or not. At this point, it is convenient to assume that each of us has a threshold level, that is, a minimum satisfaction level that we find acceptable. Below that level, when we just can't take it anymore, we decide to switch companies.[7] Since this unknown level varies from customer to customer, it is a case where uncertainty follows from the heterogeneity of preferences.

Let's put all the pieces together and formulate a very first model of why customers churn:

$$u(q - q_0) = \alpha_0 + \alpha_1(q - q_0) + \epsilon$$

In this simple behavioral model, customer satisfaction increases linearly with quality (we expect $\alpha_1 \geq 0$), and we allow for differences across customers by including a random term ϵ. Note that this is not mathematical formalism for its own sake, rather, it allows us to think clearly about heterogeneity, and our task would then be to make assumptions that describe the *shape* of the distribution here (a normal distribution if we think that this is symmetric and the tails aren't long).

---

7 In practice I am neglecting the possibility that customers compare their current satisfaction with us relative to what they expect to get elsewhere. Even if you try to get a more realistic model that includes this effect, it will be hard to have data that allows us to put it into practice. Sometimes simplification follows from the purest sense of pragmatism.

Finally, a customer will churn whenever this unobserved (or latent, as it's called in the literature) satisfaction level falls beneath a threshold level:

Customer churns if $\alpha_0 + \alpha_1(q - q_0) + \epsilon_i < k_i$

Since the threshold level k_i is also unobserved and varies from customer to customer, we may want to combine it with the random source of heterogeneity ϵ_i and just say that:

Customer churns if $\alpha_0 + \alpha_1(q - q_0) + \eta_i < 0$

This simple behavioral model allows us to build from the ground up a method to estimate our churn probabilities. Once we make a distribution assumption on our source of uncertainty (η_i), we are now ready to estimate the probability as:

$$Prob(\text{Customer churns}) = Prob(\alpha_0 + \alpha_1(q - q_0) + \eta_i < 0) = F(-\alpha_0 - \alpha_1(q - q_0)) = 1 - F(\alpha_0 + \alpha_1(q - q_0))$$

where $F()$ is the cumulative distribution function for our random variable η_i.

If you're wondering, this is how applied microeconomists estimate models of discrete choice, a method popularized by Nobel Prize in Economics-winner, Daniel McFadden and many others.

Before going on, let's discuss several points. First, you may be wondering if it's really necessary to write down everything mathematically and be very explicit about uncertainty. The answer is that the vast majority of practitioners do not go through the trouble of formalizing everything. I think it's a good practice because it forces you to think really hard about the sources of uncertainty and how to model each of them, as well as the simplifying assumptions made. Also, since you model behavior and uncertainty from the ground up, this guarantees that your estimates will be *interpretable*. This is how I usually start in my own work, and depending on time constraints and the complexity, I may proceed with a simpler approach.

A second point has to do with making the model more realistic: here we assumed that customers care only about quality, but I argued that they also care (and trade-off) about price and customer experience. Once you understand each of these and measure everything in terms of a common standard—utility or satisfaction—you can proceed and combine them together to get an aggregate measure of customer satisfaction. Note that different functional assumptions (say, additivity) imply different rates of substitution between each dimension.

Finally, we are back at the problem of estimating our probabilities. If you take uncertainty seriously, you would need to make distributional assumptions for the sources of uncertainty and proceed to estimate accordingly. For instance, if you assume normality, you end up with a *probit* model. Logistic models are appropriate if you assume that uncertainty follows a logistic distribution, and so on. Of course, first you may do some hand-waving and put everything into a black box with the hope that you get both interpretability and predictive power. In my experience this is rarely the case, though.

Cross-Selling

Figure 6-17 shows one way to analyze the problem of cross-selling different products to our customers. For each product (lever) we have to decide to make an offer that our customer may then accept or not. The problem is more interesting the more levers we analyze, but for the present discussion, we can just think of a unique lever, as that is where the main underlying uncertainty is. We will tackle the more general problem in Chapter 7, so for now, let's start by recognizing where uncertainty is found and how we may proceed to approximate it.

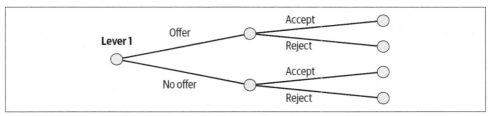

Figure 6-17. Underlying uncertainty for the case of cross-selling

At the risk of stating the obvious, customers buy products they want and can afford, and we don't really know what any of these are. The good thing about cross-selling, however, is that we already have information about our customers' previous purchase history, so we can use this knowledge to approximate the probability of their accepting our cross-selling offer. Note, however, that this alone won't suffice to estimate the probabilities: we need variation across customers who have purchased and those who haven't purchased each product.

To sum up, it is necessary that we have a sample of customers who have and haven't purchased each product. We can then try to model such choices: for instance, what else have they purchased before? Are there sequences of purchases that arise more frequently, probably because of the similarity of the products, or the type of value they create? Remember that we are trying to unravel our customers' preferences and budget constraints, and the data may reveal some interesting patterns if we ask the right questions and know where to look for the answers.

Before going to the next use case, it is good to recall that observational data may have biases that can significantly affect the quality of our probability estimates. Take the case of deciding to make a credit card offer. Since banks have historically denied access to credit to specific demographics for which they don't have enough information to make a risk assessment, most models we fit will end up reflecting exactly that low probability of "accepting." The problem here is that for these groups the offer was never made, so there was no "rejection of the offer." From an ethical point of view, the risk is creating a vicious circle where our machine learning models estimate low probabilities of acceptance for minority groups that have traditionally been denied access. From a business point of view, we may systematically be denying ourselves higher profits just because of this bias.

The solution? Before *exploiting* a lever, you may want to spend some time *exploring* other potential levers, even if costly in the short run. For instance, you may consider doing some A/B testing to circumvent some of the selection biases prevalent in observational data.

CAPEX Optimization

As we have stated the problem before, we seek to allocate investment budget across different buckets—say geographies—with the objective to get the largest possible ROI. It is customary for companies to use the whole budget, so this problem is equivalent to getting the largest possible incremental revenues.

Where is the underlying uncertainty here? Our action is how much to invest in each bucket, and the metric we seek to move is revenues. There is clearly a big gap to fill in: why would revenues increase with higher capital allocation? Since this varies across companies and across sectors, there is no unique answer to provide. But we can proceed as discussed in Chapter 5 and imagine a hypothetical scenario. Suppose that the extra capital expenditure is allocated to the improvement of our physical or digital stores. It is easier to understand now how CAPEX may affect revenues: presumably, better or larger stores can increment the volume or the ticket of the sales (otherwise, you wouldn't affect revenues directly).

In general, it must be that for each bucket something like the following takes place:

$$R(x) = P \times Q \times (1 + g(x))$$

Here, $g(x)$ is the incremental factor that depends on the size of the investment x; that is, it is a growth rate that we may want to estimate, and that may include both the price or ticket effect. As it includes both effects, and there is underlying uncertainty with each one, you may want to be more specific before going on:

$$R(x) = P\big(1 + g_P(x)\big) \times Q\big(1 + g_Q(x)\big)$$

If you're wondering what we've gained from being specific about the two effects, remember that data doesn't speak for itself; we need to ask the right questions, and in this case, at least keeping in mind that price and quantity effects may both be prevalent will allow us to look at ticket and volume data separately.

In Chapter 2 we showed that even when we understand all potential drivers of incremental revenues, selection effects are pervasive with observational data. In this case, using historical data may bias our estimates since our previous CAPEX may have been allocated in buckets with below-average performance (cities, locations, etc.), so even if there is a positive effect, our output metric may remain subpar relative to the rest of the sample of buckets.

There is no simple solution to this problem, and A/B testing may be forbiddingly costly in the case of CAPEX allocation. If this is the case, you may try finding some synthetic controls, do some matching or difference-in-difference, or some other method that is appropriate to your specific problem.

Store Locations

We are trying to decide where to open our next store, with the objective to get the largest ROI. Since our choice is the location of a store (*loc*), ideally we would want to know how profits vary with different locations, Profits(*loc*). *Without uncertainty* the first-best would be to open the stores with the largest profits.

Notice that we have done some considerable simplifying, since profits don't just appear out of nowhere when we open a store: it usually takes some time for our investments to reach a break-even (not to say the highest potential) and that time may be correlated to factors that vary with locations. So what are these factors?

To start, profits are revenues minus costs, so:

$$\text{Profits}(loc) = \text{Revenues}(loc) - \text{Cost}(loc)$$

As in the previous use case, revenues depend on the price we charge and the volume that corresponds to that price, and these, of course, vary with location. Costs may be fixed (e.g., rent) or variable (e.g., labor, electricity), and these also vary across locations:

$$\text{Profits}(loc) = P(loc) \times Q(loc) - \text{FC}(loc) - \text{VC}(loc)$$

If necessary, you can decompose the costs even further, but this level of aggregation suffices to understand what the underlying uncertainty is and the potential drivers. For instance, if you choose to open a store in a mall, other things being equal, the volume of your purchases should be higher than non-mall locations, since potential customers are attracted because of the variety of supply (malls are natural two-sided platforms or marketplaces). But even across malls you'll find variation in terms of the price you may charge, depending on the neighborhood's income level, say, among many other factors.

Now suppose that you have a dataset with historical profits for all of the stores you have opened in the past. There are two potential strategies here: you can directly estimate expected profits (the lefthand side of the last equation) as a function of factors that vary across locations, or you can estimate each of the components on the righthand side and plug them in to obtain estimated profits for different locations. It really depends on how precisely you can estimate each of the drivers of profits separately: if you can do it, this approach will be more interpretable and will force you to think hard about the underlying uncertainty and economic fundamentals of the problem.

Who to Hire

One of the hardest problems we face in our companies is the decision to hire someone, since there are many relevant uncertainties. We may simplify away the ones that are not of first-order for our current problem and consider the following three:

- Will they be productive?
- How long will they stay with the company?
- Will there be a good fit with their teams and with the company values?

We have already discussed some of the difficulties in trying to answer these questions. For instance, do we have hard productivity metrics, such as the number of sales per month or quarter? Or do we rely on softer metrics, such as a performance review done by the manager or a 360 questionnaire? Depending on the answer to these questions the demands on our data may be higher.

Let's consider the case of candidates for a sales position: the advantage here is that our productivity measure is hard, as we can directly observe individual-level performance. To take into account the first two sources of uncertainty, we may try and estimate the present value of each salesperson's profits for a given tenure, that is, their *employee lifetime value*. For now neglect the third type of uncertainty.

Why do we have such a wide distribution of sales performance across our sales force? We need to understand these factors if we want to approximate the underlying uncertainty. It could be that some salespeople have larger and better networks, or that they better understand our product, have better communication skills, or are naturally highly motivated. These and any other factors we consider important have to be assessed at the ex-ante stage, that is, before we decide to make an offer or not. Moreover, by thinking hard about it we can prepare the interviews so that we get a high signal-to-noise ratio.

Easier said than done, but we are now getting better at estimating models that allow us to predict employee lifetime values for different tasks or positions. As before, we must be *extra careful* with inherent biases that are prevalent in our data, since we want to give equal chances to candidates from different groups.

Before going on, let's talk briefly about the the third source of uncertainty: how can we assess ex-ante if a candidate will be a good fit? We should start by asking ourselves why we care about it, and presumably this is because a bad fit may either have subpar performance (this depends on motivation, and our energy levels and motivation really depend on feeling happy about our workplace, our teammates, our boss, etc), or may negatively affect the team's productivity. We may try addressing the former as we briefly discussed already—using historical data, which hopefully also includes some psychometric tests —but the latter needs a different output metric: we not only care about individual performance, but we also care about a team's *aggregate* performance. This is harder to model, and the demands on the data we need are considerably higher. Just to make this point, consider the problem of even measuring the corresponding team lifetime value: a team can go on indefinitely, but individual members may change teams or even quit and move to another company. In principle we should have data on each of these moves so that we can control for changes *within* the team.

Here's an example where we know that this underlying uncertainty matters, but because of the complexity of the task, we may decide to simplify and neglect team effects at least until we really feel comfortable with individual performance estimates.

Delinquency Rates

At the simplest level, given a loan we want to know if it will be repaid completely or not. At the next level, we may want to know how this probability changes with the size of the loan and the interest rate we charge. Note, however, that there are many reasons people will default on their debts. It could be that:

- Customer wants to pay but just can't make it
- Customer has the funds but doesn't want to pay

These are very different reasons, and in order to make an ex-ante assessment we need very different data: the former presumes honesty and the uncertainty comes from short- or medium-term shocks to the household finances (e.g., unemployment, delays in getting paid, other unexpected expenses such as healthcare and hospital charges, etc.), and the latter has to do with the underlying motivation.

Take credit scoring data: someone who has never defaulted on their obligations will have great credit history, so from this data it will be very hard to make probability assessments on their near-future family finances. Needless to say, if you've never been granted credit, you won't even have a history and will thereby continue to be rejected. Presumably, an individual's credit history may help assess the second type of motivation, assuming that this is a structural permanent characteristic: if you're the type of person who tries to take advantage of others and systematically neglects your obligations, this will be reflected in your credit score. But many times customers won't pay because they don't think it is *fair* given how they've been treated or the quality of the service they have received. A credit score will not be informative enough in this case.

Stock Optimization

We have already mentioned that the cost of overstocking is the opportunity cost of a foregone sale elsewhere or the direct depreciation cost of units being stored for a longer-than-needed time. With understocking the cost is foregone sales that could've taken place had we had enough stock in that specific location.

This discussion suggests that the main source of uncertainty we face is the volume of demand in each period. Is this enough? Let's assume that *we know* how many units will be purchased any day. Is our stock optimization problem resolved? The answer in this case is negative in general: what are the transportation costs? How much will our units depreciate? Similarly, are there risks of robbery? Notice how the trick of assuming away one specific type of uncertainty allows us to quickly identify other factors that didn't initially appear to be of first-order, but after further consideration are now important.

Key Takeaways

- *Uncertainty is everywhere*: most complex decisions we make are done under conditions of uncertainty. This means that at the time of making a decision we can't be completely sure of what outcome will result. Since uncertainty is everywhere, we had better embrace it.

- *The calculus of probabilities allows us to quantify and work with uncertainty*: we should become familiarized with the calculus of probabilities. At its core, it has addition and multiplication properties: the former allows us to compute the probability of one *or* several events taking place. The latter allows us to compute the probability of several events taking place simultaneously, and gives rise to the famous Bayes' rule.

- *The second tool we need is expected values*: since we use them all the time we must first learn to calculate and manipulate expected values. The expected value of a random variable is the weighted average of the values it takes, with weights equal to the corresponding probabilities. It's important for us because we usually calculate expected utilities.

- *To make decisions under uncertainty we compute expected utilities*: a utility function allows us to rank decisions. In many cases we take the utility function to be the *identity function* so the expected utility of a random variable like profits is just its weighted average, with weights being the corresponding probabilities. In certain applications we may wish to allow for risk aversion so we parametrize a utility function that is concave.

- *AI is our tool to quantify uncertainty*: we can use supervised regression models to directly estimate the expected utility from pulling one lever. Alternatively, we can use classification models to estimate the probabilities that each uncertain consequence arises and plug these estimates into our expected value calculation.

Further Reading

The study of uncertainty is the study of the calculus of probability. There are several introductions to probability theory out there, but Sheldon Ross's *A First Course in Probability* (Pearson) could be a good place to start. A classical reference at a more advanced level is Volume 1 of William Feller's *An Introduction to Probability Theory and Its Applications* (Wiley). Both references, as well as most introductory probability textbooks, provide a presentation of classical or frequentist probability and statistics.

If you want to delve into Bayesian probability theory, you may find Joseph Kadane's *Principles of Uncertainty* (Chapman and Hall/CRC) useful, and at the time of writing, it's available for free online (*https://oreil.ly/Yw-k_*). Alternatively, you can check out E.T. Jaynes's *Probability Theory: The Logic of Science* (Cambridge University Press). Dennis V. Lindley's "The Philosophy of Statistics" article that appeared in the *Journal of the Royal Statistical Society* provides a great discussion on foundational topics and describes in great depth the differences between frequentist and Bayesian statistics (but written from a Bayesian perspective). Different axiomatic derivations of subjective probability can be found in Peter Fishburn's article, "The Axioms of Subjective Probability," published in *Statistical Science*. Morris DeGroot's *Optimal Statistical Decisions* (Wiley-Interscience) has a fantastic presentation of this topic, as well as other advanced topics in Bayesian decision theory.

There are many great discussions online on how to interpret classical confidence intervals and p-values. You can search on Andrew Gelman's blog for many such references (for example: *https://oreil.ly/vaIam*).

On a lighter note, David Lindley's *Uncertainty* (Anchor) is a great read if you want to understand the role that uncertainty plays in science, and specifically, how quantum theory challenged all preconceived notions about it in the physical sciences. To this day, it is hard to find any other truly uncertain phenomena in the sciences. In a similar vein, Peter Bernstein's *Against the Gods: The Remarkable Story of Risk* (Wiley) provides a historical account of decision making under uncertainty, and you'll find many great references to some of the topics mentioned in this chapter.

Classical references for objective and subjective axiomatic decision theory are von Neumman and Morgenstern's *The Theory of Games and Economic Behavior* (Princeton University Press) or Leonard Savage's *The Foundations of Statistics* (Dover), respectively. You need not go to these groundbreaking pieces, however. David Kreps's *Notes on the Theory of Choice* (Routledge) is a great reference, as well as Luce and Raiffa's *Games and Decisions* (Dover). Ariel Rubinstein's *Lecture Notes in Microeconomic Theory* (Princeton University Press), which is also available for free on his webpage after registration, provides a great introduction to the topic. See also Ken Binmore's *Rational Decisions* (Princeton University Press).

Any microeconomics textbook will provide a presentation on the expected utility hypothesis, risk aversion, and many economic applications. As before, I recommend David Kreps's *A Course in Microeconomic Theory* (Princeton University Press).

Further discussion of calibration of risk aversion models can be found in Ted O'Donoghue and Jason Somerville's "Modeling Risk Aversion in Economics," which appeared in *The Journal of Economic Perspectives*, but Kadane's book referenced above has a good discussion too. On prospect theory and loss aversion, Daniel Kahneman and Amos Tversky's "Prospect Theory: An Analysis of Decision Under Risk," published in *Econometrica*, provides many experimental examples as well as an alternative solution to several of the paradoxes described here. On this same topic, you can also consult Mark Machina's "Choice Under Uncertainty: Problems Solved and Unsolved," published in *The Journal of Economic Perspectives*.

John Myles White's *Bandit Algorithms for Website Optimization* (O'Reilly) provides a great discussion on the exploration versus exploitation trade-off as well as a very applied and down-to-earth presentation of bandit algorithms. In a slightly different vein, but also applied and with code you can download and use, is Allen Downey's presentation of Bayesian statistics in *Think Bayes* (O'Reilly).

At an operational level, you may want to go beyond the classical statistical toolkit of computing confidence intervals and p-values. In that case it will be necessary to use Bayesian methods. Andrew Gelman and Jennifer Hill's *Data Analysis Using Regression and Multilevel/Hierarchical Models* (Cambridge University Press) is a good introduction; see also Gelman's *Bayesian Data Analysis* (Chapman and Hall/CRC). Kevin Murphy's *Machine Learning: A Probabilistic Perspective* (MIT Press) goes into the arduous task of providing Bayesian foundations to many commonly used methods. Finally, you may wish to search online for material on *probabilistic programming*, a recent label given to Bayesian methods that can be used at scale.

If you're interested in *microfounded* discrete choice models as we briefly discussed when talking about how to model customer churn, you can check out any microeconometrics textbook: Colin Cameron and Pravin Trivedi's *Microeconometrics: Methods and Applications* (Cambridge University Press) might be a good place to go, but Kenneth Train's *Discrete Choice with Simulations* (Cambridge University Press) is more detailed on this specific topic.

Biases in observational studies is by now an already mature area of study, and there are even startups that supposedly help companies to systematically *debias* their data. Cathy O'Neil's *Weapons of Math Destruction: How Big Data Increases Inequality and Threatens Democracy* (Broadway Books) makes a strong point on the ethical risks of using big data and algorithms that may magnify the biases in the data and have profound societal effects.

## CHAPTER 7

Optimization

After all this work, we've finally arrived at the *prescriptive* stage, the moment where we're finally ready to make the *best* decision possible. Or at least that's our aim: it won't come as a surprise that at the beginning we may make some simplifying assumptions to gain a better understanding of each problem, which can be relaxed as we feel more comfortable solving more complex versions. But first we should review some concepts from optimization theory that will be handy.

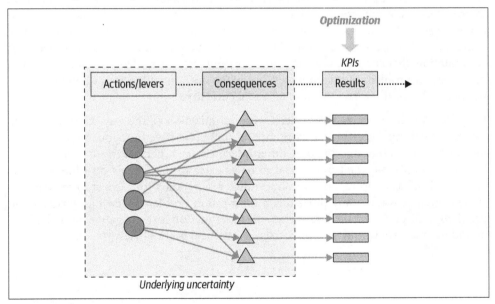

Figure 7-1. Optimization

What Is Optimization?

Optimization is about finding minima or maxima of some predefined objective function. The *objective function* is what you would've imagined: a mathematical function that maps our levers to our business objective. Since our aim is to make the best decision possible, it is almost natural that some knowledge of optimization theory will be of help.

There are cases where the problem we seek to optimize is relatively easy. Take the case of finding the maximum of two numbers: say 5 and 7. We can immediately see that the second number is larger, so if there were associated levers we would chose the second one. In any binary decision, this is all it takes in relation to the optimization stage.

If there are *finitely* many numbers, we can again sort them and readily find the maximum or minimum. It may take some time and we may not do it by hand, but there are sorting algorithms that are computationally efficient. Bear in mind, however, that a large enough, but "finite," list of numbers can be computationally expensive to sort, so we may want a more computationally efficient solution.

Some problems may look *finite* but it may still be computationally hard to find the optimum. Arguably, one of the most famous is the *Traveling Salesman Problem*, where given a number of cities (and associated distances between them) our task is to find the shortest route that goes through all of them and returns to the initial one. If you just have a handful of cities you may easily find a solution by enumerating all possible routes and finding the shortest one. But this class of combinatorial problems becomes computationally expensive once the number of cities grows.[1] This is a very common optimization problem that companies face when dealing with distribution and logistics decisions, so it's not only a mathematical curiosity.

If you're into data science, you've dealt with optimization all the way. Most, if not all, machine learning algorithms are optimizing some objective function, be it the minimization of a *loss* function or some other intermediary function. In supervised learning, for example, we want our model to be as close to the data as possible. We thus define a "loss" as a discrepancy in our model's prediction with respect to the data (in some aggregate way). Many times this minimization problem is itself a *maximization* problem in disguise, for instance, when the loss function is the negative of the loglikelihood function we would *maximize* from a statistical point of view.

---

1 If you've read about computational complexity, it belongs to the class of NP-hard problems.

When we deal with *infinitely* many values, things become much harder. Figure 7-2 starts with one relatively simple example. The left panel shows the case of *minimizing* a really nice, convex, objective function. It is common to plot the values the function takes on the vertical axis, and to denote it by *y*. Similarly, on the horizontal axis we plot the values of our decision variable or lever (*x*).

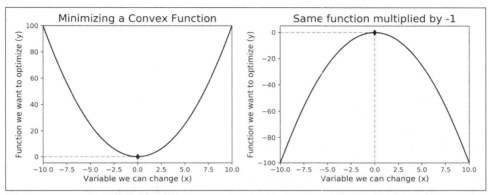

Figure 7-2. Simple optimization problem: minimization and maximization

In this simplified case, we can inspect the problem visually and notice that if we want to achieve the minimum it must be that *x* = 0. The right panel shows that maximization and minimization problems are related: if we instead want to find the maximum, we can just multiply the function by −1 and proceed accordingly. Visual inspection is no longer possible when the range of values that the decision variable takes is large enough, or when we have more than two decision variables.

Figure 7-3 shows an example of a function that is not as well-behaved since it has many local minima (squares) and maxima (stars). The word *local* here refers to the idea that you may think you've found your optimum, but once you start zooming out and see the full picture, you realize that the function keeps decreasing or increasing. The ideal for us is to find *global* minima or maxima so that we can rest assured that we've really found the very best.

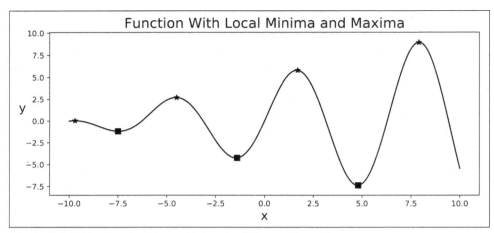

Figure 7-3. Objective function with local minima and maxima

In practice, we use computers to solve optimization problems such as the one in the previous figure. There are many algorithms that only find local minima, so we must double-check that we indeed found the optimum. There are also global optimization algorithms that search for the best solution, but these are in general hard to calibrate.

The most famous algorithm used in machine learning is *gradient descent*, where one iteratively updates the current best guess according to the formula:

$$x_t = x_{t-1} - \gamma \nabla f(x_{t-1})$$

Here x_t is our updated guess for the decision variable, x_{t-1} is the previous guess, γ is a parameter that calibrates the size of the step we want the algorithm to take at each iteration, and $\nabla f(x)$ is the gradient of the objective function we are minimizing. In the case of only one decision variable, this is just the derivative of the function, evaluated at the previous guess. If you remember your Calculus I class, you may recall that if we want to find an *interior* maximum or minimum, it is necessary that the derivative is zero at that point (this is the first-order condition).

First and Second-order Conditions for Minimization

You may recall from your calculus classes that for a differentiable objective function $f(x)$, the *first-order condition for minimization* (FOC) states that for x^* to be an *interior* minimizer, it is necessary that $f'(x^*) = 0$; that is, the derivative must be zero.

The *second-order condition* (SOC) states that if $f''(x^*) > 0$ then x^* is a local minimizer. The SOC describes the curvature of the function at a minimizer (it should be convex), while the FOC states that at an interior minimizer you shouldn't be able to increase or decrease the value of the function.

You can now see why gradient descent makes sense: first, if the derivative hasn't vanished then we haven't found a minimizer (FOC). Once it reaches zero, the algorithm will halt and no updating will take place. Also, if the derivative isn't zero, gradient descent tells us the *best direction* to keep exploring: if the derivative is positive at the current guess and we are searching for the *minimum*, it can only be that a new candidate guess is *less than* the current guess (otherwise our function would keep increasing).

Numerical optimization algorithms are very sensitive to the extra parameters, like the step parameter in the gradient descent algorithm, or the initial guesses we need to initialize the algorithm.

Figure 7-4 shows an implementation of gradient descent on the quadratic function of Figure 7-2. You can play with the parameters (initial value and step size) and check for yourself that finding the optimum for this easy problem can be problematic.

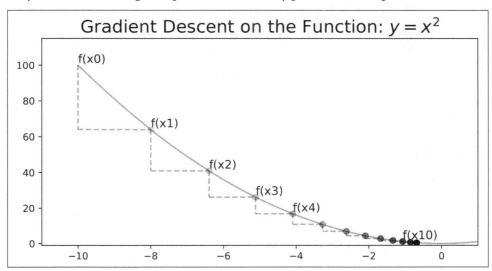

Figure 7-4. Example of gradient descent: $x_0 = -10, \gamma = 0.1$

The figure is generated with the code shown in Example 7-1.

Example 7-1. An implementation of the gradient descent algorithm

```
def gradient_descent(prev_guess, step, derivative_f):
    '''
    Given a previous_guess, the step size, and the value of the derivative_f
    this function outputs a new guess
    '''
    new_guess = prev_guess - step*derivative_f
    return new_guess

# use gradient descent to optimize the previous functions
def quadratic_fn(x):
    '''
    Example: y=x^2
    '''
    y = np.power(x,2)
    return y

def quadratic_derivative(x):
    '''
    To update we need the value of the derivative of the function
    '''
    dy = 2*x
    return dy

# two parameters we need to initialize
x_init = -10
step=0.2
# other parameters to stop the algorithm
max_iter = 1000
stopping_criterion = 0.0001
curr_diff = 100
counter = 0
while curr_diff>stopping_criterion and counter<max_iter:
    # Update guess
    x_new = gradient_descent(x_init, step, quadratic_derivative(x_init))
    # update difference, initial and counter
    curr_diff = np.abs(x_new-x_init)
    x_init = x_new
    counter +=1

print(x_new)
```

Numerical Optimization Is Hard

The advances in numerical optimization notwithstanding, doing it in practice is far from being straightforward. We need to really know well the function we want to optimize, check if it is nice or well-behaved (in the sense that there is only one maximum or minimum), play around with several initialization parameters, and be super careful about the parameters in the optimization algorithm (e.g., the size of the step in the gradient descent algorithm).

Sometimes we may really want to avoid going through the pain of doing numerical optimization at all, but depending on the use case, the benefits of doing so can be substantial. Other times, as we will see in a later example, we may find simple algorithms if we look at the fundamentals of the optimization problem.

Optimization Is Not New in Business Settings

Before data science became trendy, companies used to hire what some called *decision scientists*. They may not have been experts in machine learning, but they were really good at optimizing objective functions. Their backgrounds were in operations research, economics, applied math, or the like, and they solved very interesting and valuable problems like optimal inventories, optimal routing, or price and revenue optimization. Let's take a closer look at the latter since we have already mentioned some of this in Chapter 4.

Price and Revenue Optimization

Simplifying a bit, revenues are just price (P) times quantity (Q):

Revenues $= P \times Q(P)$

Sales depend on the price we charge—represented by the function $Q(P)$—and thanks to the Law of Demand it's usually downward sloping, creating a natural tension if we consider increasing our prices: the first term will increase one-to-one while the second one decreases as prices go up. Revenues increase as long as the first effect dominates; otherwise we are better off *lowering* our prices. Ideally, revenues as a function of price would look like those in Figure 7-5. It would then be relatively straightforward to set our prices to the optimum of $50.

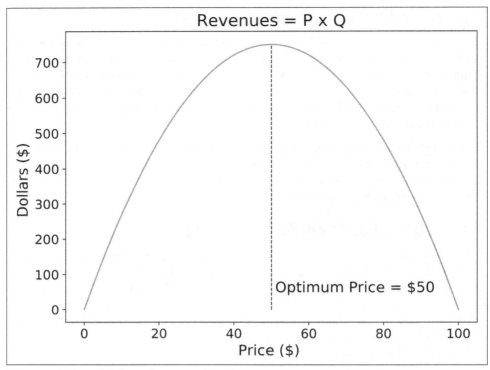

Figure 7-5. A well-behaved revenue function

What information do we need if our aim is to optimize this function? Since we don't want to be to the left or to the right of the optimum price, we need to understand how sensitive demand is to changes in prices; this is what economists call the *price elasticity of demand.*

To see this, we can just take the derivative of the revenue function *with respect to our decision variable* (prices). After doing some algebra, you can find that:

$$\frac{\partial \text{Revenues}}{\partial P} = Q(P) + PQ'(P) = Q(P) \times (1 - \epsilon)$$

where, as usual, $Q'(P)$ denotes the derivative of the demand function with respect to price, and $\epsilon = -P\frac{Q'(P)}{Q(P)}$ corresponds to the price elasticity of demand (always positive if the Law of Demand applies). We can interpret it as the absolute value of the percentage change in demand when we increase price by 1%. For example, if the price elasticity is 2, increasing prices by 1% *decreases* demand by 2%.

There is some very nice intuition behind this math: suppose you increase prices by 1%. Other things being equal, the first effect in the revenue equation would result in a 1% increase in revenues. However, since the Law of Demand applies, our sales will fall by an amount that depends on our customers' price elasticity. If sales fall by *less* than 1%, our revenues will still increase, as the negative effect isn't large enough to completely destroy the positive price increase. This shows that as long as the price elasticity of demand is *less than 1* we can increase prices and be better off. Going back to the last equation, recall that the sign of the derivative tells us the directional change of our revenues when we increase prices: since demand can't be negative, the sign depends on whether the elasticity is less than or greater than 1: in the former case, our revenues increase (positive sign); otherwise we are worse off by increasing our prices.

Decision scientists and economists alike (and some data scientists working on this type of problem) spend most of the time estimating the price elasticity of demand for their products when their aim is to set optimal prices. This is not the place to delve into the difficulties of this enterprise, but I will provide references at the end of the chapter.

Optimization Without Uncertainty

As already mentioned, *optimization can be very hard*, but we can start by simplifying our problems by assuming away all uncertainty. Let's see some examples to understand the power of simplification when doing optimization.

Customer Churn

Figure 7-6 shows the case of customer churn without uncertainty. In this ideal scenario, we would know the true state of our customer: they may leave with absolute certainty, they may stay with us for sure (for now, things may change in the future), or they might be considering changing, but if we make a good retention offer, they will stay. Since we have the magic eight-ball, we also know how good that retention offer has to be (the minimum to make them stay).

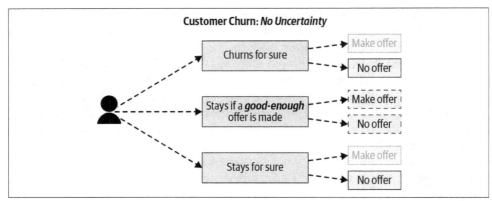

Figure 7-6. Customer churn: the case of no uncertainty

Note that in the case of the pure states (leaves or stays with absolute certainty), we should *not* make a retention offer: in the first case, there's an opportunity cost (time, our team's effort designing the strategy, not targeting better candidates) and we should target other customers instead. In the second case, our customer will most likely accept the offer, and that has a direct cost to us (e.g., foregone revenues if it takes the form of a price discount) as well as a similar opportunity cost as in the first case. As shown in Figure 7-6, in these two cases we are always better off by not making the retention offer.

The intermediate case is more interesting: since we know the minimum offer we need to make (there's no uncertainty, remember), for the business case to work it must be that the incremental revenues are at least as large as the cost we incur by making the retention offer. There are customers where a small enough offer will suffice and others where it's just not profitable for us to try to retain them.

Solving the problem with no uncertainty helps us fine-tune our choice of levers and intuition, but also allows us to clearly identify where the fundamental underlying uncertainty is for the problem at hand.

What Are We Optimizing in the Case of Customer Churn?

Let us clarify what objective function we seek to optimize in the case of no uncertainty. Recall that we may be willing to make a costly offer (at a cost c) if we generate an *incremental* revenue of v (representing the stream of future revenues if the customer stays with us).

Then, for each type of customer we have:

If customer churns for sure
> The customer will churn no matter what offer we make, so profits are:

$$\text{Profits} = 0 - c = -c$$

> Since we always have a loss, it is never optimal to make the offer.

If customer stays for sure
> The customer will keep purchasing from us anyway, so incremental profits are also negative:

$$\text{Profits} = 0 - c = -c$$

> Again, it's never optimal to make an offer.

If customer churns depending on the offer
> For a customer that accepts and stays, we make:

$$\text{Profits} = v - c$$

> Since there is no uncertainty, we know for each customer the minimum offer needed to be accepted. The optimal rule for this type of customer is to make an offer only if it will be accepted and we make non-negative profits.

Cross-Selling

Figure 7-7 shows the case of cross-selling without uncertainty. In this case, we have four levers we can pull (for now let's consider only that our levers are different products we may offer). What should we do?

To highlight where the main underlying uncertainty is, the figure already shows what the customer will do when we offer each product. Products one, two, and four will be accepted, each with positive incremental value. Product three will be rejected, so once we take into account the cost of making the offer (say, because of marketing spend), the return becomes negative.

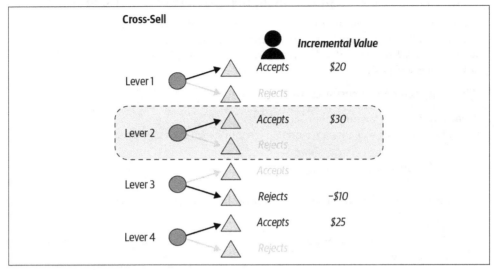

Figure 7-7. Cross-selling: no uncertainty

If we assume the customer only buys one product, which one is the *next-best offer*? It takes no time to see that we should offer the second product, since that's the one that maximizes the incremental value. This is clearly an oversimplified example, but it shows the main trade-offs we face: with products that will be rejected, we have a direct cost and an opportunity cost of not offering the *right* product. With products that will be accepted, are we offering the one that maximizes our return? If not, there's an opportunity cost (foregone earnings) that our competitors will happily accept.

You can start making this problem *without* uncertainty more complex. If the customer is willing to buy more than one product, should we offer them together, or even bundle them, thus reducing the overall price of the combo? It really depends: if a competitor is able to offer one or all (at the same prices, say), then we face an opportunity cost and bundling is now a viable alternative (if regulation permits). If the order matters, say because products are ever more complex, then we may decide to wait (as this constraint also applies to our competitors).

The important thing to remember is that making an offer that is unlikely to be accepted has true costs: there are direct costs such as marketing spend and those associated with creating and contacting the leads, but also real opportunity costs that are usually dismissed. One that is often neglected is information overload: we just keep sending emails to our customers, and after a while they just systematically dismiss them, thereby closing a valuable sales and communication channel.

CAPEX Investment

Let's go back to our CAPEX investment example, where we have to decide how much to spend in several different buckets. One such example is when we need to invest in different cities, so for the rest of this example, we will be talking about geographical units.

As we have assumed in earlier chapters, the effect that capital expenditures have on each city's revenues is:

$$\text{Revenues}(x) = P \times Q \times (1 + g(x))$$

where $g(x)$ is the growth rate in sales due to higher investment. Since we are in the realm of no uncertainty, we can take this as known. Two alternative ways to model the growth rate are depicted in Figure 7-8, where I have parametrized growth using a generalized logistic function of the form:

$$g(x) = A + \frac{K - A}{\left(C + Qe^{-Bx}\right)^{1/v}}$$

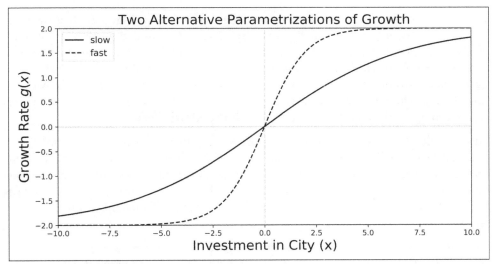

Figure 7-8. Alternative growth rates

The figure is generated with the code shown in Example 7-2.

Example 7-2. Assumed logistic growth for CAPEX optimization

```python
def logistic_growth(x, A,K,C,D,B,nu):
    '''
    Generalized logistic function
    '''
    return A + (K-A)/(C + D*np.exp(-B*x))**(1/nu)

# generate plot
fig, ax = plt.subplots(figsize=(10,5))
x = np.linspace(-10,10,100)
y1 = logistic_growth(x, A=-2,K=2, C=1, D=1, B=0.3, nu=1)
ax.plot(x,y1, color='k', label='slow')
y2 = logistic_growth(x, A=-2,K=2, C=1, D=1, B=1, nu=1)
ax.plot(x,y2, color='k', label='fast', ls='--')
ax.plot([-10,10],[0,0], color='k', alpha=0.2)
ax.plot([-10,10],[2,2], color='k', alpha=0.2, ls='--')
ax.plot([-10,10],[-2,-2], color='k', alpha=0.2, ls='--')
ax.plot([0,0],[-2.1,2.1], color='k', alpha=0.2, ls='--')
ax.axis([-10,10,-2,2])
ax.set_title('Two Alternative Parametrizations of Growth', fontsize=18)
ax.set_xlabel('Investment in City (x)', fontsize=18)
ax.set_ylabel('Growth Rate $g(x)$', fontsize=18)
ax.legend(loc=2, fontsize=12)
```

Using logistic functions to model growth has several advantages: first, growth is bounded by the asymptotes, in this case, by –2 and 2. Bounded growth is not only realistic (at least in the short term), it will also bound total revenues, making the optimization problem feasible. It is also smooth, so we will be able to take derivatives everywhere without a problem (and also, in this case we can find the derivatives analytically, which always speed up our computations). This is important if we are using a computer to do the hard work of finding optimal allocations.

Also, we allow for the negative effect of disinvesting in a city, since the decision variable can take negative values. Consider the case of closing stores in one city to open additional stores in another city. This disinvestment can have pervasive effects on that city's revenues, and our model allows for this. The important thing is that *it may be optimal* to disinvest in some locations, and the model is general enough to allow for this scenario.

But another property is readily seen in the figure that will drive some of the results. Note that at the beginning, when we start spending in a city, the growth rate accelerates very quickly, but at some point *decreasing returns* start kicking in, so every additional dollar we spend has an ever-decreasing effect (until it completely disappears).

The figure shows two alternative parameterizations where decreasing returns kick in at different rates. In the first case (*slow*), it takes a larger investment for the city to reach its maximum potential growth. In the second case, growth accelerates very fast at the beginning and then stalls.

Under these assumptions, total revenues are just the sum of revenues over all cities. In the case of two cities, our optimization problem becomes:

$$\max_{x_1, x_2} \text{Total Revenues}(x_1, x_2) = P_1 Q_1 \times (1 + g_1(x_1)) + P_2 Q_2 \times (1 + g_2(x_2))$$

subject to $x_1 + x_2 = \text{CAPEX}$.

Note that in principle, growth rates could vary across cities, so I label them accordingly. Also, this is a constrained optimization problem, where we have a total budget (CAPEX) that needs to be allocated across cities. Without further (positivity) restrictions, I also allow for *disinvestment* in each city.

Starting with two cities allows us to understand the fundamental trade-offs of the optimization problem, and we will also be able to visually inspect our objective function. Note that since the budget has to be exhausted, we can make this two-variables problem one that depends only on the investment in one city, say x_1. This is done by replacing $x_2 = \text{CAPEX} - x_1$ in the objective function.

Figure 7-9 shows what our objective function looks like in the symmetric case where the two cities are exactly the same in terms of size and purchasing power (so the average ticket, or price in our revenue equation, is the same, and I take initial sales to be the same too), and also in terms of the growth rates. I plot two different parameterizations of these growth rates, where I only change the speed of acceleration as in Figure 7-8.

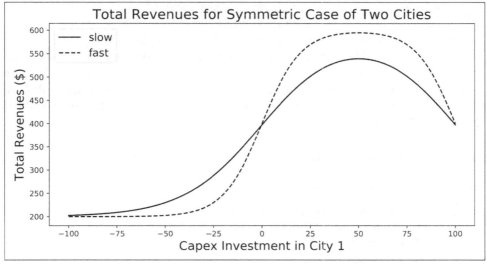

Figure 7-9. Total revenues for symmetric example of two cities

The symmetric case is a good place to start, since it allows us to test our intuition (and our code, of course). Here I take total budget to be $100 and plot total revenues as a function of investment in city 1, where we can exhaust the total budget or disinvest the same total magnitude ($-100 \leq x_1 \leq 100 = \text{CAPEX}$).

As the figure shows, since both cities are exactly the same, we should expect the optimal allocations to be the same too (why would we overinvest in one?). In both cases you can verify that indeed the optimal investments are $x_1^* = x_2^* = 50$.

We could use numerical optimization, but in this case, the economics are simple enough that we can construct an algorithm that will easily generalize to many cities. The idea is as follows: suppose you only have $1. You can either assign it to city 1 or 2 (you can't split it as this is our fundamental unit of accounting). Where should we spend it? Simple economic intuition suggests that it should go to the city where it creates larger incremental (marginal) revenue (since the incremental cost is the same in both cases and equal to $1). So we proceed accordingly, and assign it to the better city. If incremental revenues are the same, it doesn't matter, so just assign it to the first one, say.

Imagine that our CFO now gives us an extra $1 to allocate. The intuition is the same here: we should assign it to the city with higher incremental revenue, but notice that something has changed: *we have already assigned the previous dollar* to city 1. If growth rates display decreasing marginal returns, the extra dollar will have lower incremental revenues in city 1, and depending on the actual growth rates, these may or may not be larger than those for the second city.

The pseudoalgorithm works like this: we start allocating extra dollars to the city that wins in incremental returns at each iteration, update current total investment in each city, and proceed until we exhaust our budget. I like this algorithm, not only because it is easily generalizable to many cities and forces the constraint to be satisfied, but also because it captures the fundamental economic reasoning all the way (and by making some small changes it will also allow for optimal disinvestment) so that we can easily explain it to our business stakeholders. The following Python snippet (Example 7-3) shows one implementation of this algorithm:

Example 7-3. Iterative solution to the CAPEX allocation problem

```
def revenues_in_city(x,**kwargs):
    '''
    Compute revenues as a function of investment (x)
    Note: keyword arguments (**kwargs) allow us to parametrize growth
    (A,K,C,Q,B,nu) and price, sales
    '''
    P,Q,A,K,C,D,B,nu = kwargs['P'],kwargs['Q'],kwargs['A'],kwargs['K'],
                       kwargs['C'],kwargs['D'],kwargs['B'],kwargs['nu']
    revenues = P*Q*(1+logistic_growth(x, A,K,C,D,B,nu))
```

```
        return revenues

def find_the_optimum_iteratively(CAPEX, **kwargs):
    '''
    We can assign one dollar to the city with largest incremental
    (marginal) revenue
    '''
    # to allow for asymmetric cases we can include some parameters
    p1,q1 = kwargs['P1'],kwargs['Q1']
    p2,q2 = kwargs['P2'],kwargs['Q2']
    B = kwargs['B']
    # Initialize array to save incremental allocations
    optimal_assignments = np.zeros((CAPEX,2))
    for x in range(CAPEX):
        # compute revenues for both cities if we assigned the marginal dollar
        # 1. Accumulate previous investments (if any)
        x_init1, x_init2 = np.sum(optimal_assignments[:x,:], axis=0)
        # Test allocations: current allocations plus an additional $1
        x_test1, x_test2 = x_init1+1, x_init2+1
        # Compute Marginal Revenues: rev(x+1)-rev(x)
        marg_rev1_x = (revenues_in_city(x_test1,P=p1, Q=q1, A=A, K=K, C=C,
                       D=D, B=B, nu=nu)-
                       revenues_in_city(x_init1,P=p1, Q=q1, A=A, K=K, C=C,
                       D=D, B=B, nu=nu))
        marg_rev2_x = (revenues_in_city(x_test2,P=p2, Q=q2, A=A, K=K, C=C,
                       D=D, B=B, nu=nu)-
                       revenues_in_city(x_init2,P=p2, Q=q2, A=A, K=K, C=C,
                       D=D, B=B, nu=nu))
        #print('Iteration ={0}, rev1 = {1}, rev2={2}'.format(x,
              # marg_rev1_x, marg_rev2_x))
        # if they are the same, it doesn't matter where you assign it
        if marg_rev1_x==marg_rev2_x:
            optimal_assignments[x,0] = 1
        elif marg_rev1_x>marg_rev2_x:
            optimal_assignments[x,0] = 1
        elif marg_rev1_x<marg_rev2_x:
            optimal_assignments[x,1] = 1
    # when done, we get the full trajectory of investments
    return optimal_assignments
```

Figure 7-10 shows the trajectories of investing marginal dollars in our two-cities symmetric example. Note that at the end we have reached the symmetric solution: both cities get assigned $50 as the previous figure had shown should be the case.

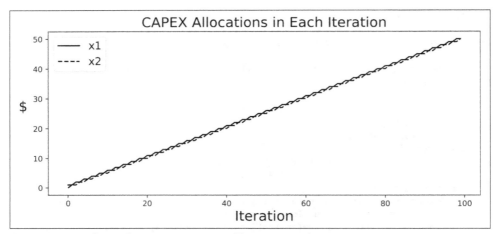

Figure 7-10. Solving the optimization problem iteratively

Once we feel comfortable with our solution in the simplest, symmetrical case, we can move forward and tackle the case of asymmetric cities. Say that both cities have the same purchasing power, but one is 10x larger than the other (proxied by sales). What do you think will happen?

The left panel in Figure 7-11 shows total revenues where everything is the same for both cities, except the size (Q). We can now see that the optimum is biased against the smaller city (city 1 in the figure), as the marginal returns of investing in the larger city are also larger. Interestingly, we *do not* allocate all the budget in that city. This is because marginal returns eventually kick in, and each extra dollar in that city is worth less to the company. You can see what happens in each iteration in the right panel: the first $72 are allocated in the largest city, at which point marginal revenues have fully kicked in.

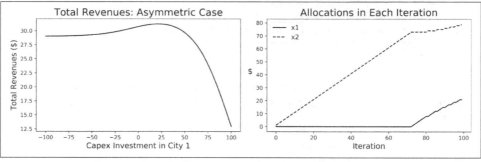

Figure 7-11. Total revenues and allocations in the asymmetric case (city 2 is 10x larger)

You may wonder why, if the second city is 10x larger, optimal allocations are not 10x larger too. These are the types of questions that our business stakeholders will ask, and we will only be prepared to answer if we really understand the workings of our model. Notice that the question implies some type of linearity, but our growth rates are everything but linear: not only are they bounded, but decreasing returns to investment are built into our model from the start.

Before moving on, notice that this type of optimization problem abounds in business settings. If you are in marketing, for example, you may want to decide the optimal allocation across different channels (this is the famous marketing mix modeling). If we know the growth rates or ROI functions for different channels, the optimization solution follows along the same lines. We now need our data scientists to make use of their extraordinary machine learning skills and estimate these functions for us.

Optimal Staffing

Even though it has been stressed throughout, in the prescriptive stage the choice of the objective function is crucial, and it should be tied directly to our business objectives. This will be pretty clear in the case of optimal staffing.

As the name suggests, we want to find the optimal number of employees to maximize some business objective. In the abstract, we can think of many such objectives, for instance, our company's profits. Casual observation suggests that most companies tend to hire *more* people relative to this optimal level, so wouldn't it be great if we could optimize the size of our workforce? The answer, of course, is yes, but the truth is that this is a very hard problem, since it's difficult to get precise, trustworthy employee-level productivity measures. If we were able to do so, we know what the optimal rule would look like: we should hire an extra employee if her impact on our revenues is larger than her cost.

So let's propose a different objective, one where it's relatively straightforward to get individual-level productivity measurements. Consider the case of finding the optimal number of cashiers we should have in our stores. Their productivity can be measured by the number of customers served per unit of time, so the metric that is most directly affected is the time customers wait in line. We may later wish to tie this to our business performance via its effect on our customers' satisfaction—which itself affects the churn probability—but for now let's take it as our objective.

Let's start with one store. Figure 7-12 shows a high-level picture of the queueing problem: customers enter at a rate λ (measured in terms of customers per period), and each cashier services μ customers per period, so if we have n cashiers, each period $n\mu$ customers exit the store. The length of each period is defined by us and is crucially restricted by operational constraints: we may be able to solve the staffing problem for periods of five minutes, say, but it is usually infeasible to make changes at such short time windows.

Notice that other than the size of our staff (our decision variable), these quantities are generally uncertain: entry depends on demand and commonly shows highly seasonal patterns (e.g., peak hours, lunch breaks, weekends, etc.), and service rates depend on each cashier's productivity, which not only varies across our staff, but also for each individual throughout the day. Since we are simplifying away all uncertainty first, we will assume that these are fixed and known.

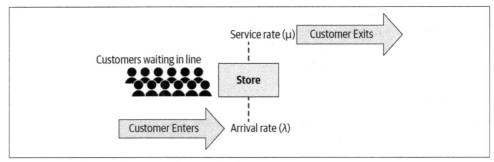

Figure 7-12. Queue theory: preliminary concepts

As usual, before solving the problem without uncertainty we need to make some simplifying assumptions: let's start by assuming that customers form in one line and are served on a first come, first served basis. This is how we design our queue. Also, customers don't leave the store even when the queue is long (they are infinitely patient), and they all arrive at the beginning of each period, so that if we take periods of one hour, say, all customers in the period arrive on the zeroth minute.

Before proceeding, let's discuss the validity and impact of our assumptions. The implementation details of our queue are our choice, and without uncertainty they won't play a role; that is, we may instead have one different line for each cashier and the analysis that follows will remain true. Assuming that customers are infinitely patient is done for mathematical convenience: it's just easier to count the waiting time for customers that wait. If they left, not only would the size of the queue change (and the waiting time for those after them), but it would also be hard to compute the average waiting time as we would have to decide what to do with those whose time was truncated because of early departure.

The last assumption—that customers arrive at the beginning of the period—will play an important role, and will certainly *overestimate* waiting times in our analysis. The reason I'm making it is that it's easier to count waiting times if the clock starts at the same time for everyone (I can use the common clock); otherwise I would have to start the clock for each customer once they enter. An alternative assumption would be that they enter uniformly throughout the period, so that a constant fraction of customers enter as time goes by: 6 over the first 10 minutes, 12 over the first 20 minutes, and so on. I will come back to this assumption as we see some of the properties of this queueing model.

Let us start with a very concise example, where each period denotes an hour. Suppose that the entry rate is $\lambda = 100$, so that 100 customers enter the store per hour. Also, each cashier serves 20 customers per hour ($\mu = 20$), implying that it takes 3 minutes to serve each customer (this is the reciprocal $1/\mu$).

Figure 7-13 shows several important properties of queues when we consider having $n = 4, 5, 6$ cashiers. Starting with the top panel, consider the case of having four cashiers. Since everything is deterministic, we know the waiting time for each and every customer that entered during this 60-minute batch, each one represented by a circle. The intensity of the shade represents how long they wait in line.

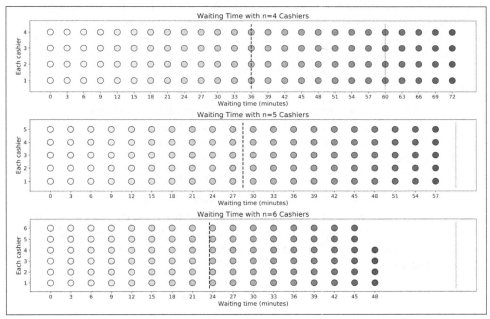

Figure 7-13. Queues with three alternative choices of staff size (n = 4, 5, 6): each circle represents a customer, and its shade denotes the time they wait in line (lighter means less time)

Note that the first four customers (from left to right) in line are served immediately, so they have to wait zero minutes. It takes exactly 3 minutes for each one of the four cashiers to finish serving their corresponding customer, then the next batch of four customers is served. The next four will wait in total 6 minutes, and so on until the 100 customers that entered during the period are served. If we average the waiting time across all customers, we get that they wait 36 minutes (vertical dotted line), and one-fifth of the customers actually had to wait for *at least one hour*, the exact time a new batch of 100 customers enters the store (vertical solid line).

This last result—batches that overlap since the previous one cannot be fully served during the period—holds whenever $\lambda > n\mu$, so queues will keep on accumulating. When $n = 5$ (middle panel) we can see that the whole batch of 100 customers is fully served during the period, and average waiting time is now 28 minutes and 30 seconds. If we include the 3 minutes that customers have to wait, we see that when $\lambda = n\mu$, as is the case here, the whole 60 minutes are taken serving all customers. In this case, queues don't accumulate without bound and the system actually finds an equilibrium or steady-state (when queues keep accumulating, waiting times grow without bound).

The bottom panel shows the case when we have more than the minimum number of cashiers for the equilibrium to be reached, $n = 6$. Average waiting times keep decreasing (to more than 23 minutes) and as before, everyone is served within the period. Note in passing that as a result of having an extra cashier, there is now some dead time in each period that we may wish to include in our cost-benefit optimization analysis.

To sum up, when we have less than the minimum number of cashiers for the queues to be finite, the waiting time is unbounded. This is because when the second batch arrives in the top panel in the figure, the first four customers in that second batch already have to wait 15 minutes to be served, in total, so you can see what happens with every succeeding batch: the store would literally never close.

Going from 4 to 5 cashiers has huge marginal returns, as measured by the waiting time. Going from 5 to 6 cashiers reduces the waiting time by almost 5 minutes. Figure 7-14 shows this pattern for up to 50 cashiers (when average waiting time reaches 1.5 minutes). Example 7-4 shows the code. We see that it pays off to hire more cashiers, or does it?

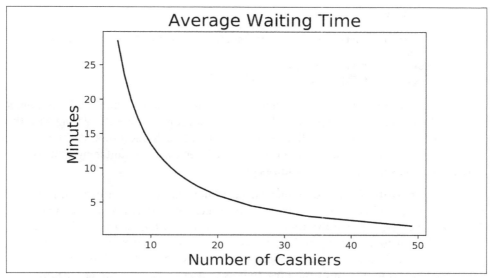

Figure 7-14. Average waiting time decreases with the numbers of cashiers

Example 7-4. Computing waiting time in deterministic staffing problem

```
def compute_waiting_time_deterministic(entry_rate, service_rate, n_cashiers):
    '''
    Given entry_rate, service_rate, n_cashiers
    Simulate waiting time in a deterministic setting
    '''
    # Get a DF to compute average waiting time
    # in the deterministic optimal staffing problem
    n = n_cashiers
    time_per_customer = 1/service_rate
    # save results in a DF:
    # rows: groups of n customers served at a time
    number_of_rows = int(np.ceil(entry_rate/n))
    df_simu = pd.DataFrame(n*np.ones((number_of_rows, 1)), columns=['customers'])
    # check that total customers in line match the entry_rate
    cum_sum = df_simu.customers.cumsum()
    diff = df_simu.customers.sum()-entry_rate
    df_simu.customers.loc[cum_sum>entry_rate] -=diff
    # compute waiting_time per group of n customers served
    df_simu['wait_time'] = df_simu.index*time_per_customer
    # Average will be waited by each group's share
    df_simu['frac_waiting'] = df_simu.customers/entry_rate
    # Share x Waiting Time of each Group
    df_simu['frac_times_time'] = df_simu.frac_waiting*df_simu.wait_time
    # accumulate to find weighted average (last observation will be average
    # waiting time)
    df_simu['cum_avg'] = df_simu.frac_times_time.cumsum()
```

```
        return df_simu

results_df = compute_waiting_time_deterministic(entry_rate=100, service_rate=20,
        n_cashiers=5)
avgwt = results_df.cum_avg.iloc[-1]
print('Average Waiting Time (minutes) = {0}'.format(avgwt*60))
```

It really depends on whether the incremental savings are larger than the incremental cost. As the figure shows, incremental savings in waiting time are *decreasing* with the number of cashiers. If on the other hand the cost of each additional cashier is, say, constant, our maximization problem will be well behaved, guaranteeing that we can find an optimal staff size that maximizes profits.

Let's go back to the assumption that all customers arrive at the beginning of the period. We already know that if entry is distributed (say, uniformly) along the period, then we are *overestimating* the waiting time for some of the customers. To see this, think of the last four customers that were served when we have six cashiers (bottom panel in Figure 7-13). Under our assumption, they waited 48 minutes until served, but in reality, assuming uniform entry, they would've entered in the last 10 minutes, *when there was no one in line*, so their waiting time would have been zero instead.

Recall that solving the problem without uncertainty serves the purpose of refining our intuition about the fundamental forces driving the results, and as such we could just leave it here and proceed to solve it under uncertainty. If you still want to get a better picture you can do one of several things. First, you can try estimating the average time we overestimated (say, under the assumption of uniformity). But this would defeat the purpose of solving the problem without uncertainty. Second, you could try breaking the intervals into smaller ones that make sense from your business standpoint. When the equilibrium condition holds ($\lambda = n\mu$), queues in a batch are independent of those in other batches (since all customers will be served within the period), so we can proceed and solve the problem in periods of 30 minutes, say. Third, you can implement individual clocks for each customer (as real stores do today with their queuing systems).

Finally, we need to map waiting times to business performance. In a world without uncertainty mappings are relatively straightforward, at least conceptually. Waiting time matters from a customer experience perspective, and unsatisfied customers will have no problem buying from our competitors, thereby reducing our future revenues. If we had at hand such a rule that converts average waiting time into churn rate, we could compare the cost of each additional cashier with the incremental saved revenues, making the optimization problem solvable.

But this is only one possible objective function. To highlight the importance that the objective function has, let's imagine that we don't want to maximize profits, but we'd rather have a homogenous customer experience *across all stores*. Since for any store we already know how average waiting times map to the number of cashiers, our optimization algorithm could instead redistribute from the top performer in the region to the worst performer, one cashier at a time, until convergence. The metric we seek to minimize here is the difference in waiting time between the worst performer and the top performer.

Optimal Store Locations

Suppose we are considering opening stores, given our fixed yearly budget that we intend to exhaust. Under no uncertainty, and if we already have a fixed set of candidate locations denoted by their geographical coordinates (lat, lon), the problem is straightforward: we just order in descending fashion all locations according to their stream of future profits, including set-up costs, and either exhaust the budget or stop with the store that breaks even, whichever happens first (Figure 7-15).

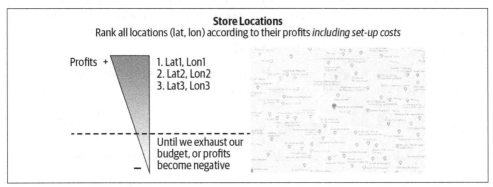

Figure 7-15. Optimizing where we open new stores

This was an easy problem without uncertainty because we started with a (finite) pre-filtered number of candidate locations. The more general problem where we don't prefilter is much harder since there are possibly infinitely many locations (just think of the number of possible coordinates you see on any map), but the general principle of sorting locations still applies. You may want to make things finite by dividing the map into a rectangular grid and having the average long-term profits of each rectangle as the metric to sort.

Optimization with Uncertainty

As we saw in Chapter 6, an optimal decision under uncertainty is obtained when we maximize our expected objective function, so, in principle, "all we need to do" is replace our objective function with its expectation. Sometimes this is relatively straightforward, but other times it isn't, and we will need to find ways to still solve our problems, at least to a first-order approximation. In any case, it is always recommended to try to understand the problem without uncertainty as this will signal exactly where you should focus your attention (not all uncertainties are created equal).

Customer Churn

We already saw what the fundamental sources of uncertainty are in the case of customer churn: whether the customer stays or leaves *without* an offer being made, and the minimum offer we would need to make for those that are undecided.

Denote by $p(x)$ the probability that the customer will churn given an offer of value x. I would expect the probability of churn to be decreasing with this value: in the extreme case of giving away a new car, say, most customers will stay with us (all of them *should* stay with us, at least in the short term, since no matter how much they despise our company they can still profit from selling the car). $1 - p(0)$ corresponds to the pure case of those customers that will remain loyal to the company *without* the need to incentivize them, and for a very large x—$\lim_{x \to \infty} p(x)$—we find those customers who leave no matter how large the offer.

For now, fix the values of the offer to two possible values, say $x = c$ or $x = 0$. The expected return of making an offer to a customer of CLV v is:

$$E(\text{return} \mid \text{offer} = c) = (1 - p(c))v - c$$

We make an offer whenever our expected profits are larger than for the case of not making the offer:

$$(1 - p(c))v - c \geq (1 - p(0))v \Longleftrightarrow v(p(0) - p(c)) \geq c$$

The right-most condition is just a reminder of the general rule when we are doing optimization: make an offer whenever the incremental saved revenues (in expectation) exceed the incremental cost.

This is a very nice inequality as it prescribes to which customers we should make an offer and which should be let go. To make it practical we need our data scientists to use their ML toolkit and estimate the churn function $p(x)$ (and I assume we have customer-level data on CLVs). We can then find the break-even customer where the inequality becomes an equation.

Note how setting and solving the prescriptive problem takes us very far from the predictive state of affairs. Most companies ask their data scientists to estimate ever-more predictive churn probabilities. But as this optimality condition shows, we ought to go one step further: we need them to estimate the conditional probability of changing companies *given* the size of an offer we make. And this is not an easy task for several reasons.

The first one is that there is uncertainty due to heterogeneity of our customers' preferences. We can imagine that the minimum acceptable offer is smaller for some customers and larger for others. We may be satisfied at the beginning to have an average estimate.

But most importantly, the problem of estimating this probability remains. As we have discussed before, observational data is less than ideal to do so. The recommendation here is to do some experimentation: choose two or three (or several) offers of different value, randomize across your customers, and estimate their impact on the probability. By conditioning on some of our customers' characteristics (e.g., age, gender, tenure, history of transactions, etc.) we can take care of part of the heterogeneity, taking us one step further in our quest to full customization.

Cross-Selling

As we already saw, in its purest form, assuming that we know the incremental value of each possible product we are considering selling to each of our customers, all we need to do is estimate the probability that a customer will accept our offer $Prob(\text{Accept}|X)$, possibly by conditioning on some customer and product characteristics X.

For each customer we would then just rank the expected value of each offer and design our campaigns in such a way that we always make the *next-best offer*, that is, the one that maximizes the expected value.

This is "all" it takes to build next-best offer models. Of course, the devil is in the details, and it takes a lot of practice to estimate these probabilities accurately.

Basic economics tells us that our customers will buy a product if they want it (preferences) and they can afford it. So we would like to condition on prices (this takes us back to "Price and Revenue Optimization" on page 151) and characteristics that proxy our customers' preferences. This is easier said than done, but we need to start somewhere and iterate forward with more sophisticated models.

Optimal Staffing

As we saw earlier, the two fundamental parameters in the optimal staffing problem are the entry and service rates, so it won't come as a surprise that we need to make distributional assumptions about them. One such probability model is the *Erlang C model* where the entry rate follows a Poisson distribution and exit or service rates follow an exponential distribution. As in the case of no uncertainty, it is also assumed that customers are infinitely patient and wait in line. All of these assumptions are questionable from an empirical point of view, so depending on your application you may want to resort to more advanced stochastic models.

With n cashiers, the store's average utilization rate is denoted by $\eta = \lambda / n\mu$. It can be shown that under these distributional assumptions, the probability that a customer has to wait in line takes the following form:

$$P_C = \frac{\frac{(n\eta)^n}{n!(1-\eta)}}{\sum_{i=0}^{n-1} \frac{(n\eta)^i}{i!} + \frac{(n\eta)^n}{n!(1-\eta)}}$$

We can then proceed to calculate the expected waiting time as we did in the case of no uncertainty, or calculate stricter customer service metrics, such as the *grade of service* (GoS), defined as the probability that a customer has to wait at most a fixed *acceptable waiting time* (AWT):

$$GoS = 1 - P_C e^{-\mu * (n - A) * AWT}$$

It may not be immediate from these formulas, but the intuition we developed from solving the problem without uncertainty suggests that the probability of waiting *decreases* with the number of cashiers, and the GoS *improves* as our staff grows larger. Consider the case of a store that receives an average of 100 customers per hour. The service rate is 3 minutes per customer, and we want to find the number of cashiers needed for the grade of service to be at least 80% with an AWT of 1 minute. Figure 7-16 shows the Erlang C probability of waiting and the corresponding GoS as we increase the number of cashiers. From the right panel, we see that we need 7 cashiers to achieve this GoS.

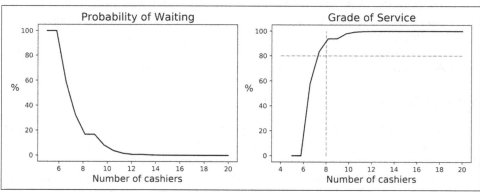

Figure 7-16. Waiting probability and grade of service

With these expressions at hand we can formulate different prescriptive problems. For instance, we can proceed as in the CAPEX example and pose the problem of finding the number of cashiers that maximizes our incremental profits:

$$\text{Profits}(n) = P \times Q \times (1 + g(GoS(n, AWT))) - c(n)$$

The underlying assumption here is that the GoS affects customer satisfaction (and churn), impacting future sales and revenues. We can have lower churn (higher revenues) by hiring extra cashiers, but only if we incur the corresponding labor costs. Crucially, note that we have excluded all other costs when we pose this problem (rent, electricity, etc.) since these do not change with staffing decisions, at least to the first order.

At this point it's good to review what we have done: we modeled uncertain entry and service rates by making strong distributional assumptions that allow us to derive analytically an expression for the GoS, or the probability that a customer waits a fixed acceptable waiting time. What we need now is a mapping between the churn probability and GoS.

Here you can do one of several things. You can parametrize the growth rate function to depend positively on GoS, and try to estimate or calibrate the corresponding parameters. There is a bit of hand-waving here, since we are no longer maximizing *expected* profits (unless everything is linear), but it might be OK to sacrifice probabilistic accuracy and have a problem that we can solve.

Alternatively, you can look at the microfundamentals of churn and derive a function that depends directly on AWT. For instance, we may follow a route similar to that in Chapter 6 and say that a customer churns once their waiting time reaches a maximum threshold or limit, that is, the point where they just can't take it anymore and decide to switch companies.

Conveniently, we can call this *individual-level* threshold their own *acceptable waiting time*. It follows that the probability that a customer i churns depends on their own awt_i:

$$\text{Prob}(churn_i \mid awt_i) = \text{Prob}(\text{time } i \text{ waits} > awt_i) = 1 - GoS(n, awt_i)$$

Unfortunately we don't know each of our customers' awt_i, but we suspect there is heterogeneity as some customers will be more patient and others will quit sooner. The important thing is that we have already found a mapping between the (negative) growth rate in sales to GoS derived from the probabilistic model. We can then plug in reasonable values of the AWT and solve the corresponding optimization problems, possibly with a sensitivity analysis.

Tricks for Solving Optimization Problems Under Uncertainty

The last example took us very close to the following general problem. Suppose we want to maximize the expected returns that depend on some random variable θ:

$$\text{Max}_x E(f(x; \theta))$$

Don't get lost with the abstract notation here: θ is our fundamental source of uncertainty (the random variable), and x is the decision variable. For instance, θ can be the waiting time for each of our customers, and x the numbers of cashiers we hire.

What I want to highlight is that unless the problem is relatively simple, this objective function might be a complicated object: recall that the mathematical expectation can either be a sum of finitely many terms involving probabilities, or an integral in the case of continuous random variables.

One trick is to simplify as much as you can (but not too much). For instance, try to solve the problem with binary random outcomes. This is what we did with the churn examples: a customer either leaves or stays. The nice thing about using threshold levels to binarize decisions is that we can translate a model with infinite heterogeneity into one that separates the universe into two camps: those that are above the threshold and those that are below.

Another trick is to use *linear* objective functions since the expectation operator is linear. Let's look at the revenue example again:

$$f(x) = P \times Q \times (1 + g(x))$$

If the growth-rate function $g(x)$ is linear in x, say $g(x) = a + bx$ for some constant values a, b, we can just plug in the expectation of the random variable x. Note that this *won't* be the case if we use a logistic function as in the CAPEX optimization example.

The important thing to remember is that in general it *won't* be the case that $E(g(x)) = g(E(x))$, unless the function $g(x)$ is linear. We are getting way too technical, so I'll just leave it here with that warning.

One final alternative to computing difficult integrals is to do a Monte Carlo (MC) simulation where we draw many random variables from a chosen distribution, compute the objective function for each draw, and average all of these computations. The *chosen* distribution must match the source of uncertainty in our decision. The intuition here is that the sample average of the function (across our draws) approximates well the expected value.

To see how this works, consider the example in Figure 7-17 where we seek to solve the following minimization problem:

$$\text{Min}_x E(\theta x^2) = E(\theta)x^2 = 10x^2$$

with θ a normal random variable with mean 10 and variance 36 ($\theta \sim N(10, 36)$). In this case, our objective function is linear in the random variable so we *don't need to do simulations*. My objective here is to showcase its use in a very simplistic setting with the hope that you get the main ideas.

The simplicity of the example notwithstanding, note that since the random variable can change the sign of the second derivative, the shape of the function might change drastically depending on the realization.

The left panel in Figure 7-17 shows the distribution that our random variable follows, and highlights the shaded area corresponding to the probability (~4.7%) that we end up minimizing a *concave function*. The right panel shows 100 realizations of our objective function: in almost 95% of the cases we end up minimizing a convex function (with the optimal decision being zero), but in the remaining cases we might end up looking for a minimum with a concave function.

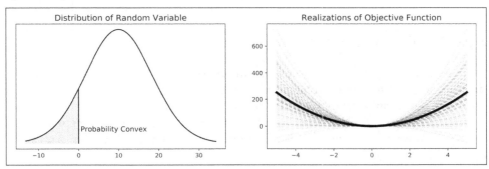

Figure 7-17. Simple case of stochastic optimization

In a nutshell, the idea behind MC simulation, as applied to this type of optimizaton problem, is that under relatively standard assumptions, our best estimate of the population expectation $E(g(x; \theta))$ is the sample average computed from the observed N realizations of the random variable:

$$E(g(x; \theta)) \sim \frac{1}{N} \sum_{i=1}^{N} g(x; \theta_i)$$

Using the code in Example 7-5, we obtain the following result, with 100 draws from the corresponding normal distribution (Figure 7-18). The dotted line is our true objective function ($10x^2$), and the solid line is the result from our MC simulation.

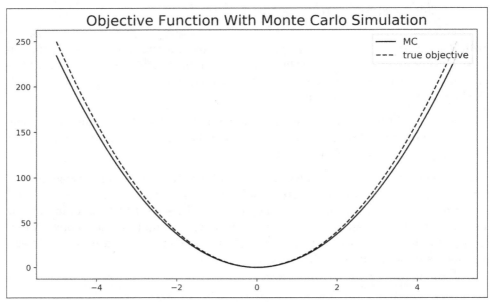

Figure 7-18. Example of Monte Carlo simulation

Once we have the objective function, we can proceed to find the optimum just as in a deterministic optimization problem. It's a bit trickier since we don't have a function anymore, only a finite set of numbers. But we already know how to solve this type of optimization numerically (some variation of sorting), or if the problem is high-dimensional we can fit a parametric function over this simulated dataset that can be optimized.

Example 7-5. Simple Monte Carlo simulation

```
def quadratic_stoch(x,epsilon):
    return epsilon*x**2

# Monte Carlo simulation
def get_montecarlo_simulation(objective_func,grid_x, N_draws):
    '''
    For each realization we will have (x_k,objective_func(x_k,realization))
    We then average _across_ realizations
    Inputs:
    objective_function: pointer to a well-defined Python Function that takes
    an array (grid_x) and returns an array of the same size
    grid_x: the grid over our decision variable where we wish to evaluate
```

```
the function
N_draws: size of our Monte Carlo simulation
Output:
An array of length, len(grid_x) of sample averages across draws
'''
# get random draws: in this case we know its normal N(10,36)
draws_from_normal = 10 + 6*np.random.randn(N_draws,1)
# Initialize matrix to store results
K = len(grid_x)
matrix_results = np.zeros((K,N_draws))
for i,draw in enumerate(draws_from_normal):
    eval_i = objective_func(x=grid_x,epsilon=draw)
    # save in corresponding row
    matrix_results[:,i] = eval_i
# Find sample average:
sample_avg =  np.mean(matrix_results,axis=1)
return sample_avg
```

Key Takeaways

- *The prescriptive stage is all about optimization*: since we want to make the best decision possible, we need to be able to rank all options and choose the optimal one.

- *Many problems are simple*: when we have only two (or just a few) levers or actions we can pull, we just need to be able to calculate the expected utility, sort them in a decreasing fashion, and choose the top alternative. So the hard part is computing the expectation and sorting.

- *But in general optimization is hard*: when we have many (an infinite) possible actions, optimization becomes hard. Most likely we will use numerical methods —like the gradient descent algorithm used in machine learning—to find the solutions, and these are difficult to calibrate, *even with simple enough problems*.

- *Always start by solving the problem with no uncertainty*: solving this simpler problem will provide valuable insights and intuition as to what the relevant underlying uncertainty is, as well as the workings of levers and objective function. We can then move forward and try solving the more general problem.

- *A clear statement of the prescriptive problem will help our data scientists understand exactly what needs to be estimated or predicted*: as in the case of customer churn, to solve the prescriptive question we may need to ask our data scientists to estimate objects that are quite different from what we usually ask from them. In that example we don't need an estimate of the probability that a customer will churn, but rather how this probability varies with the size of the retention offer.

Further Reading

There are many great references in optimization theory, for instance Stephen Boyd and Lieven Vandenberghe's *Convex Optimization* (Cambridge University Press) or on a more introductory level, Rangarajan Sundaram's *A First Course in Optimization Theory* (Cambridge University Press). Applied numerical optimization is covered in Jan Snyman and Daniel Wilke's *Practical Mathematical Optimization: Basic Optimization Theory and Gradient-Based Algorithms* (Springer), or at a more advanced level in Kenneth Judd's *Numerical Methods in Economics* (MIT Press).

The best reference I know on *Pricing and Revenue Optimization* (Stanford Business Books) is Robert Phillip's book with the same title. *The Oxford Handbook of Pricing Management* (Oxford University Press), edited by Özalp Özer and Robert Phillips, is a great general reference on the topic of pricing, covering everything from specific strategies in selected industries to the theoretical foundations of pricing. A more informal reference that provides good use cases is Jagmohan Raju and Knighton Bliss's *Smart Pricing: How Google, Priceline, and Leading Businesses Use Pricing Innovation for Profitability* (FT Press).

The Erlang C model is very standard in queueing theory, so you can find references in any book on stochastic processes (see references in previous chapters).

Finally, on Monte Carlo simulation, you can check out Christian Robert and George Casella's *Monte Carlo Statistical Methods* (Springer). A good review article on the use of these methods to solve stochastic optimization problems that also provides other references is the one by Tito Homem-de-Mello and Güzin Bayraksan found online (*https://oreil.ly/121Rl*).

## CHAPTER 8

Wrapping Up

In the last few chapters, we covered a lot of material. The analytical toolbox is so diverse that many of these skills may seem disconnected. Let's finish by zooming out and taking a look at how the pieces fall into place. We'll also discuss some difficulties we may encounter and finish up with my own view on how things might evolve in the future.

Analytical Skills

In each of the preceding chapters we covered at least one skill. Let's wrap up by putting everything in perspective and discussing some challenges that remain ahead.

Asking Prescriptive Questions

We first need to learn to ask business questions, and specifically, we need to learn to ask *prescriptive* business questions. To do so we first have to learn to distinguish this type of analysis from its descriptive and predictive counterparts, and this is something that takes practice.

The good thing is that we are always making decisions, so we can learn and practice this skill by continuously looking at the choices we make, both in the workplace and in life. Our analytical muscle is like any other muscle, and we just need to be constant and conscious about exercising it.

Here's a summary of the type of questions you can ask if you want to practice and further develop this skillset:

- Why are we making the decision in the first place?
- What is the problem we want to solve?

- Why do we care about this problem, and what are the outcomes we care about?

- What are the actions we can take to attain these outcomes?

- How have we been doing in the past?

- What is the ideal dataset we need to have a granular picture of our actions and their relation to the outcomes?

- What are the consequences of our actions, and what do they depend on?

- What is the underlying uncertainty, and how would we go about making this decision without uncertainty?

These questions apply to any decision we make, so just start working with your most basic day-to-day decisions and try to answer each of them.

Descriptive analysis is the most natural for us, but doing *good* descriptive analysis is a skill we must also develop. When done correctly it can be enlightening, as it helps to redirect any initial theories of why things work as they do and show where the causal traps are that we may encounter. A good rule of thumb that most advanced practitioners apply is to start with the question and move backward to search for answers in the data.

And of course, you must start with the *right* questions. As I've said before, it's an understatement that identifying the right questions is a skill as important as learning the workings of machine learning algorithms and how to apply them. No matter how good the algorithm is and how great your data looks, starting with the wrong question necessarily leads to a dead end.

Predictive analysis by itself doesn't generate value, unless our decisions were already optimal or very good to begin with. Predictions are inputs to our decision process, and current machine learning and AI technologies are here to support us.

Asking prescriptive questions will force us to:

- Focus on the right metrics or objectives we wish to impact
- Question and enlarge the current set of levers we have at our disposal

How do we learn to ask prescriptive questions? I believe that we must start by questioning what goals we are pursuing. That's essentially what the sequence of *why* questions does.

Understanding Causality

Once you've settled on the business objectives, you can then focus on the process to achieve them. You want to be sure that you ask the following questions:

- What are our current levers?
- Why and how do they work?
- What theories or assumptions are needed?

Asking these questions necessarily takes us to the problem of causality. *Understanding* causality is different from *testing* for causality, of course. The former has to do with our ability to create theories that connect our actions with our business objectives. "By pulling this lever, that consequence will follow, thereby affecting our objectives." But why does the outcome follow? What are the hidden and not-so-hidden assumptions?

Disentangling causality empirically is usually very hard, as we saw in Chapter 4. Before investing in acquiring that skill, it's good to learn to *question* what we believe are causal findings. Can we come up with counterfactual explanations? Are there any other variables that may explain the result once we control for them? Are there selection effects?

Answering these types of questions is easier than estimating causal effects. The quick way to do this is by embracing the experimental method, that is, by systematically doing A/B testing. The most data-driven companies run thousands of experiments per year just to have a strong foundation to make better decisions. If experimenting is not possible because it's too costly, for instance, there are a myriad of statistical methods to test for causality. For these to work, your data needs to satisfy one or several assumptions, so it usually takes some expertise to learn and apply these methods.

But as soon as you start making and testing theories you will find that the world may work very differently from what you'd planned. Just as an anecdote, the 2019 Nobel Memorial Prize in Economic Sciences was awarded to three researchers that embraced the experimental method in their study of poverty and economic development and failed at it many times (*https://oreil.ly/WF9sW*). So be prepared to fail, over and over.

Thinking outside the box

Failing over and over will force you to think outside the box. It is usually the case that we start with the most obvious choice of levers to test. Then, when things don't work as planned, we start questioning our most basic understanding of our business and human nature.

It sometimes happens, however, that the more we develop our analytical muscle, the harder it is to think creatively. Analytical reasoning tends to be quite linear and systematic, which may feel far away from the ability to think creatively. We must always try to remember that this could be happening. Having a strong experimental culture may help mitigate these effects. But running cross-functional teams with a diversity of skills should also help.

Simplify

No matter how powerful our brain is, the workings of our businesses—not to mention the world—are just unsurmountable to understand all at once. It would be great if we didn't have to simplify, because then we would be able to make the absolute best decision.

In this sense our solutions are generally *local* optima. This just means that most likely there's some better decision we could've made, but solving that more general problem is just too complex right now.

That's fine. Our objective is to beat our competitors, and their businesses are also run by humans, no matter how smart they are or how much computing power or data they have. We're all in the realm of finding local optima. Start simplifying, solve the problems at hand, and iterate to the next level of complexity if it's cost effective.

Embracing Uncertainty

All of our decisions are made under uncertainty. This matters because when we pull a lever or make a decision, we can't be sure about the outcome that will follow.

We should always start by solving the problem under the assumption that there's no uncertainty. This will not only help us refine our intuition about the levers and key underlying uncertainty—the one that is of first-order to our problem—but will also serve as a benchmark to compare the quality of our decisions under uncertainty.

This said, there are at least three approaches to deal with decisions under uncertainty:

- Do nothing
- Use a data-driven, brute force approach
- Use a model-driven approach

Do-nothing approach

You may decide that dealing with uncertainty is too costly. Learning to work with the calculus of probabilities is in no way easy. It takes time to internalize its logic and make it intuitive. But all in all, the benefits are generally larger and you can always hire someone with this skillset. Our task in the enterprise is to create teams with individuals that have complementary skills.

Since the do-nothing approach is generally suboptimal, it makes sense to invest some time and money to get the right skills for our companies. Moreover, current AI and machine learning advances make it relatively cheap to find those skills: many people are continuously learning and improving the data scientist toolbox, and the price of (greater) computer power has gone down by several orders of magnitude. If anything, the cost of dealing with uncertainty has never been cheaper than it is today.

The data-driven approach

Many companies follow this path: they have the data, invest in the right technology, and hire the talent. The underlying principle of this approach is that data by itself will show us how to deal with uncertainty. But data alone won't tell us anything. Or even better, data can tell us so many alternative stories that we won't know which one is most suitable for our use cases.

To deal with uncertainty—as with creating theories about why some levers work—we need to make assumptions, to create models.

The model-driven approach

The model-driven approach makes strong, falsifiable assumptions about the sources and workings of uncertainty. We start by identifying each of these sources, classify them, and focus on the ones that are of first-order to our problem.

The next step is to make distributional assumptions about the relevant underlying uncertainty. This could be done by looking at the data. Is it uniformly distributed? Does it follow the normal bell shape? Does it have long tails?

Alternatively, we may start at the drawing board *without* looking at the data and think hard about the sources of uncertainty. Since most uncertainty isn't pure—like quantum uncertainty—but rather the result of our ignorance, heterogeneity, our need to simplify, or the result of complex interactions, we should always start by understanding each source and think about whether we can make distributional assumptions to take care of it.

Think about flipping a coin: in the data-driven approach we look at the outcomes of repeating many times the experiment of flipping a coin, and estimate that on average heads and tails are equally likely to result. But we may also start by seeing that with a fair coin, there is nothing that should bias the outcome to either side. Here we start from a symmetry assumption, come up with a simplified theory, and check our theory with data.

If it's still unclear how the model-driven approach works, think about the solution proposed to the *optimal staffing problem* in Chapter 7. After acknowledging that the main sources of uncertainty are the rate at which customers entered and exited, we made strong assumptions about how these are distributed.

With both the frequentist and Bayesian approaches we make distributional assumptions that need to be checked with our data, but the latter makes the process of thinking hard about uncertainty more explicit.

Tackling Optimization

Prescriptive analysis is all about optimization. If we are aiming at making the *best* decision possible, we have to be able to rank alternative options.

But optimization is hard. Let's review some of the difficulties we usually encounter.

Understanding the objective function

The objective function maps our actions or levers to the business objective or metric we want to optimize. We should always start by thinking hard about this objective function:

- What does it look like? Graphing more than two actions or decision variables can be hard, but it's usually good to plot the objective function against each of the decision variables *leaving all others constant* or fixed at some value.
- Is it "smooth"? Can you take derivatives anywhere? This will matter if you plan to use numerical, gradient-based methods.
- Does it have many local optima?
- Even if it is well-behaved, is it relatively flat at the optimum or on some other range? This might impact the speed of convergence of any algorithm you use.

As with uncertainty, we can approach our derivation of objective functions in model- and data-driven ways. The data-driven approach can be performed as any other supervised learning task: our features are our levers or actions—and any other controls we believe are important—and the outcome to be predicted is our business objective or metric. Needless to say, to use this method, we need to observe both the levers and business objective and have enough variation in our levers to provide reliable estimates. Moreover, you may need a supervised method that allows you to estimate non-linear effects to make the problem more realistic.

There are two problems with the data-driven approach. First, nothing guarantees that the results will make sense from a business standpoint. This happens because we generally don't impose any business structure or restrictions on the prediction problem in a supervised learning algorithm. Correlation does not imply causation, and while predictive technologies most generally capture correlations, optimization is about causation from actions to objective.

Moreover, the resulting objective function—the predicted values from actions to business obective—may not satisfy any desirable properties for optimization. For instance, if you are doing maximization, we may want to have a concave objective function. And nothing in the predictive algorithm can guarantee this to be the case.

To see this, think about price and revenue optimization (Chapter 7). Depending on our exact business objective, the data-driven approach would estimate a predictive model of revenues, profits, or contribution margin that depends on prices (and other controls) and use this predicted relationship as our objective function. Alternatively, we could start by *modeling* our objective function by thinking about the economic fundamentals: revenues equal price times quantity, which itself depends on price via the Law of Demand, and so on. By putting structure on the problem we can guarantee that it both makes business sense and also has desirable properties.

Dealing with local optima

We may have found a solution to our optimization problem, but how do we know this is the best we can do? This problem is especially true with numerical optimization algorithms that tend to settle down when no further improvements can be made. We first need to check that indeed our solution is a minimizer/maximizer depending on our business problem, and then try with different initial guesses or solutions to see if we can find other optima.

Sensitivity to initial guesses

As mentioned previously, one way to take care of multiple local optima is to try different starting points to initialize our optimization algorithms. This lack of robustness in finding numerical solutions is quite standard, and we should always check that our solution is robust to alternative initial guesses.

Scaling and production issues

The problem of scale is whether the time and complexity of solving an optimization problem increases more than proportionally as more data comes in. This is a well-known problem in machine learning, and many times we choose to productionize scalable algorithms that may not be state of the art from a predictive perspective.

Let's briefly discuss some common trade-offs:

- If new or more data comes in, should we solve the optimization problem again?
- If we do, can we react fast enough? That is, can we solve the optimization problem in a timely fashion?
- If we can't, how sensitive will our optimal recommendations be to new or more information? How often should we solve the optimization problem again?

These questions are not at all easy to answer. For instance, if our resulting optimal actions are *mappings* or functions that depend on other information, we may simply reevaluate such mappings on the new information, which may not be computationally expensive.

To see this, think of optimal pricing for an airline. In such a scenario, it is not uncommon that the optimal pricing rule depends on the time until departure (time difference between booking and taking the flight). If we know what this function looks like, we can simply evaluate the function and set optimal prices at any given time. On the other hand, if we need to solve the optimization problem every time we want to set prices, this may be hard to scale or put into production.

The AI-Driven Enterprise of the Future

What does the AI-driven company of the future look like? While I'm not in the business of making futuristic predictions, it is still worthwhile to discuss the opportunities available to us thanks to *current* technology. To this end, let's start by restating the three main points of this book:

- We create value at the enterprise by making decisions.
- Prediction is an input to making better decisions.
- There are other skills that are complementary to predictive technologies and thus necessary to improve our decision-making performance at the enterprise.

Back to AI

Current AI are predictive technologies, backed by ever-more powerful machine learning algorithms that need tons of data and computing power. Of course, AI is much more than this, but the current hype is mostly about deep learning algorithms. The natural question is whether we can make better decisions with our current technology. One avenue proposed here is to use predictions from machine learning algorithms as inputs in well-defined decision problems, and specifically, to use them to embrace uncertainty.

But can we use it in other ways? Most successful companies that have embraced the AI paradigm have become very good at translating business problems into predictive problems that can be solved with the current technologies. Can we translate any of the decision-making stages into predictive problems?

Learning how to make decisions

Let's imagine a hypothetical scenario where we translate the decision-making problem into a predictive one. The output variable is the outcome of the decision, as measured by our business metric or objective. Alternatively, we could transform it into a classification problem where every decision is labeled as either *Good Decision* or *Bad Decision*.

Now, machine learning algorithms learn from data, so let's continue to imagine that we also have a dataset of decisions made in the past. Not only do we have the output variable as described previously, but also the levers or actions that we pulled (and by how much they were pulled). Since decisions are made in different contexts, we would also need as much contextual information as possible to be used as inputs. Contextual information would not only allow us to interact our decisions with the context, but also to approximate uncertainty.

Mathematically, our supervised learning algorithm would try to learn the mapping from decisions D to business outcome y, given some contextual information X:

$$y = f(X, D)$$

In principle, under these conditions, we could just train a machine learning algorithm to make these decisions. If the outcome variable is a continuous business metric such as *Profits*, after training the model we could just turn on and off each possible lever in our dataset (each of the dummy variables in the matrix D), simulate the expected profits, and choose the one that returns the maximum. If we had transformed the output variable to some categorical labels like *Good Decision* or *Bad Decision*, we would repeat the simulation of pulling the levers and stay with the one with the maximum probability score.

Let's take the case of self-driving cars where we *automatically* store every decision made by a human driver and all of the available contextual information captured by all sorts of sensors. We could then use this approach to make driving decisions. The nice thing about the self-driving cars example is that the dataset could be created automatically, so it serves as a good benchmark for the approach.

Some problems with this approach to automatic decision-making

The first problem is creating the dataset as most decision problems we encounter are not as easy to store as the self-driving car example. But even if we could create such a dataset, it may not have the desirable properties needed for causal extrapolation.

Recall that machine learning algorithms create predictions from correlations, and as such, are powerful pattern-recognition devices. Ideally we would systematically randomize our choices or levers to different contextual information, but this, of course, may be costly and even dangerous.

Moreover, any bias in the data will be reflected in the decisions made by such algorithms, which may raise serious ethical concerns.

Finally, we have the problem of interpretability. Powerful learning algorithms are black boxes where it becomes very difficult to explain why a prediction was made. In our case, since predictions become decisions, the problem is even greater.

If you have read elsewhere the problems that current AI face, you may recognize them immediately in this list. Prediction technologies used to automate our decisions will necessarily inherit all of the shortcomings of the current technology.

Ethics

I briefly touched upon the ethical dimension in the previous section, but this is a topic that deserves some discussion in and of itself.

From the point of view of AI-driven decision-making, the main source for ethical concerns is to be found in the predictive stage, where we use vast amounts of data to make predictions. When data is biased, machine learning predictions will most likely reinforce that bias. Let's see how this process works.

Suppose you want to predict acceptance of a new credit card for a bank, with the purpose of making a decision to lend or not. Since most banks have traditionally been conservative, many underrepresented demographic groups for which they do not have enough credit data are either blocked from the outset or they are allowed to enter the funnel but are eventually denied credit. A machine learning algorithm very quickly finds a pattern: "underrepresented minority potential customer is denied credit." It doesn't go deeper and it certainly doesn't make a cost-benefit business case.

Naturally, if we use the results from the predictive algorithm we will end up denying credit to the affected demographic group, reinforcing the already biased credit decision. And this has nothing to do with credit risk. It's all about the nature of the data, and the problem of how to correctly debias data, at scale, is being actively researched in the machine learning community. It can also impact the company through bad press.[1]

Another concern that is less discussed—as it's more difficult to scale—involves situations in A/B testing where we randomly assign one treatment to one group and the alternative to another control group. For academic researchers in many universities around the world, whenever they want to run an experiment with human subjects they have to get approval from an institutional review board or ethics committee. While most companies don't require committee approval for testing, one should always remember that testing has consequences for individuals.

---

1 For a recent example, see "Apple Card Investigated After Gender Discrimination Complaints" (*https://oreil.ly/mIj-B*) in the *New York Times*.

Consider the case of displaying two different layouts on your webpage. One group of randomly selected customers will view the traditional layout, and another group will view a new version that you want to test. The outcome you want to measure is the conversion rate, that is, how many of the viewers end up buying from you. This simple setup can raise ethical questions: are you negatively affecting one group? Of course, if you just try a different color the concern may be unfounded, but you can imagine many realistic scenarios where it's less clear-cut (for example, showing pictures of abused children or war victims).

The problem with ethics is that it's hard to differentiate between something that is morally wrong or not in a precise, clear-cut manner. To see this, think of a cell phone manufacturer that wants to estimate the impact that bad hardware or software has on their revenues (via customer churn). Hardware malfunction is hard to predict and quality control is costly, but we may want to find the percentage that is optimal for the company, say 3% or 4%. The company decides to run the following A/B test: produce and randomly sell malfunctioning cell phones to a group of customers, and measure the impact it has on customer churn relative to the control group. If the test is correctly designed it will provide a clear-cut estimate of the causal effect of hardware malfunction and customer churn, and we can then go back and find the optimal level of quality control we want.

How do you feel about this experiment? I've discussed several variations of the experiment with colleagues and students, and most people find it disgusting, even when the story includes a committed customer support service to guarantee that ex-post the customer is as good or better than one in the control group.

To close, ethical concerns abound when using some of the tools in the analytical toolbox and we must, at least, be conscientious about them, and try, as much as we can, to debias any outcomes that can affect specific people or groups.

Some Final Thoughts

We have come a long way since the big data revolution started with its promise of generating ever-increasing profits for our companies. Thanks to it and to advances in computing power the AI revolution came along, and we're just beginning to understand how much can be done.

The main takeaway from this book is that value is created by making decisions, not by data or prediction. But data and prediction are necessary inputs to make AI- and data-driven decisions.

To create value and make better decisions in a systematic and scaleable way, we need to improve our analytical skills. I have described the ones that I've found most useful as a practitioner and decision-maker at different companies. I've also found that students, business people, and practitioners alike are hungry to understand better ways to create value from these new technologies, so my hope is that this book takes them in that direction.

Can we expect decisions to be automated in the near future? In some sense, many simple decisions have already been automated, as in the case of simple environments where nothing changes. But it is unlikely that human-like complex decisions will be automated by some type of Artificial General Intelligence in the near future. Until that happens, there is plenty of space for us to create more value for our companies and society as a whole.

A Brief Introduction to Machine Learning

In this Appendix I briefly summarize what machine learning is with the purpose of providing a self-contained guide. I have not attempted to go into the intricacies of the methods, as the topic of the book is to learn to create value from these technologies, and not to learn each of these many different methods. It will provide some background knowledge and hopefully some intuition on how machine learning works. For interested readers, I will also cover the basics of A/B testing.

What Is Machine Learning?

Machine learning is the scientific discipline that studies ways that machines learn to accomplish certain tasks by using data and algorithms. Algorithms are recipes or sequences of instructions that are applied over and over until a precise objective is attained, and are written with programming languages that enable human interaction with computers. These are then translated to machine language that can then be processed and computed.

A Taxonomy of ML Models

Any taxonomy of machine learning algorithms starts by describing *supervised* and *unsupervised* methods (Figure A-1). In general, learning is supervised when someone or something tells us when the task was completed successfully. For instance, if you are learning to play a musical instrument, say the conga drums, your teacher might show you first how a good slap tone sounds. You try it yourself and she tells you whether the technique and sound were close to a perfect tone; we call this process of comparing an attempt with the ideal case *supervision*.

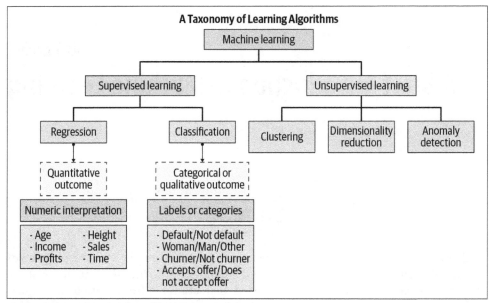

Figure A-1. A taxonomy of learning models

Supervised Learning

Supervised learning algorithms work the same, with humans providing guidance on how the world is and how different it is from the algorithm's current guess. For this to happen we must first *label* the data. Take an image recognition example where we input a picture and the algorithm needs to correctly identify what appears in the picture (say dogs). We feed the algorithm with enough pictures of dogs with the "dog" label, as well as pictures of cats, drums, elephants, and any other object in the world with their corresponding "cat," "drum," and "elephant" labels. Thanks to labeling, the algorithm is able to compare the most recent prediction with reality and adjust accordingly.

The supervised world is relatively straightforward; all we need is a method to generate predicted labels, a method to compare these with the real ones, and an updating rule to improve performance over time. We evaluate the overall quality of our current guesses by means of a *loss function*: it achieves a minimum whenever our data is perfectly predicted and increases as the predictive power worsens. As you might imagine, our objective is to minimize the loss, and smart updating rules allow us to keep descending until it is no longer feasible to get significant improvements.[1]

---

1 The most commonly used updating rule in ML is the *gradient descent* algorithm we showcased in Chapter 7.

Unsupervised Learning

Unsupervised learning is much harder since there's no explicit measure of what a right or wrong prediction is. Our objective is to uncover some underlying pattern in the data that is also informative for the problem at hand. Two examples are clustering and anomaly detection.

In clustering, our aim is to find relatively homogenous groups of customers, say, where the different groups are also different from each other. This is commonly used in data-driven segmentation.[2]

In unsupervised anomaly detection, the algorithm tries to distinguish "expected" from "unexpected" by looking at the distribution of observed characteristics. For instance, it may be that 95% of our executives sell between 7 and 13 computers in a regular day; anything off these limits could be labeled as abnormal.

Semisupervised Learning

There is an intermediate category not depicted in Figure A-1, sometimes called *semisupervised* learning, where an algorithm is able to generalize knowledge from just a few examples of data. Some practitioners believe that this is closer to how humans learn, in sharp contrast to the most advanced supervised techniques that have to be fed with thousands or millions of data points to provide reliable predictions. Children learn to recognize objects from just a few examples. Some researchers believe that this semisupervised learning problem is one of the most important challenges that current AI must overcome.[3]

Regression and Classification

To continue with our taxonomy of learning algorithms, let us talk about the two most general classes of supervised learning tasks: regression and classification. In regression our labels take the form of quantitative measures. For instance, you may want to predict the age or income level of your customers, or the time we believe they will be loyal to our company; all of these can be accurately represented as numbers: your 40-year-old neighbor is one year older than your 39-year-old brother and seven years younger than your 47-year-old friend. In regression, *values have a strict numeric interpretation*.

---

2 This is in contrast to business-driven segmentation where the analyst provides explicit business logic to group the customers. For instance, one group could be males between the ages of 25 and 35 with tenures with us longer than 24 months.

3 This category is also called *self-supervised* learning. See *https://oreil.ly/yky7W*.

In contrast, in classification tasks the objective is to predict a categorical label: you may want to predict your customers' gender, if they will default on their loan or not, or whether one specific sale is fraudulent or not. Notice that the labels here are categories and you can assign them any name you want. In gender classification, "men" can alternatively be labeled "0" and "women" labeled "1," but these apparently numeric labels lack numeric meaning: you can just switch the labels and redefine everything accordingly and nothing has changed in your learning task. There is no sense that one is larger than the other or you cannot perform arithmetic calculations.

Regression can be easily converted into a classification problem by taking advantage of the ordered property of numbers. One common regression task in business is to predict the profits from a specific activity, but sometimes you may not need the exact numerical value; say you are happy to know if your company is going to lose money (negative profits) or not (zero or positive profits). Similarly, for advertising and marketing purposes we usually need age ranges associated with behavioral differences, instead of an exact estimate of a customer's age. In both examples we end up with classification problems instead of the original regression tasks. Note, in passing, that this is the same trick advocated in Chapter 7 to binarize decisions where a continuum of actions can be taken.

Notice that while we can easily bucket any numbers into ordered groups, reverting the process to recover the original numeric labels may not be feasible. We might be tempted to approximate these by means of some statistic such as the average value in the category, but this is generally not recommended. As an example, take your customers' schooling level, accurately measured as years of schooling. Imagine that for your business problem, you only care about predicting if your customers completed college or not. We can now use the college threshold and divide the world in two: a customer's schooling level is either higher than completed college, or not. Note that later, if you try to revert the labeling, you may run into trouble as the strict numeric interpretation in regression might be broken. And most certainly, you should always avoid arbitrary relabeling of each category (say 1 for completed college and 0 otherwise) and using this in a *regression setting*, as the regression algorithm will literally treat the labels as numbers (and not metaphorically as wanted).

Making Predictions

Since current AI are prediction technologies, it is useful to get a sense of how prediction comes about. To this end let us start with a rather abstract, but general enough description of the task. Our objective is to predict an outcome y that we, humans, conjecture depends on inputs (also called features) $x_1, x_2, \cdots, x_K$. To get a prediction we need to transform inputs into outcome, which in mathematical terms can be described as a function $y = f(x_1, x_2, \cdots, x_K)$. For instance, we may want to predict our next quarter's business profits (y), and conjecture that they depend on the severity of the weather (x_1) and our projected labor costs for the quarter (x_2). In this case we only have two features, so Profits = f(severity, labor).

At this point, it is unclear where this function comes from, but notice that if we had such a function, prediction should be relatively straightforward: we just plug in the values of our inputs on the right-hand side of the equation and the output is our prediction. For instance, we can evaluate our prediction function for weather severity of 100 inches of rainfall and labor costs of \$15,000 as $f(100, 15000)$. If we had a linear function $2,000x_1 - 2x_2$, we plug in the values and get that the profits are $2,000 \times 100 - 2 \times 15,000 = \$170K$.

Caveats to the Plug-in Approach

There are two caveats to the simplicity of this approach. First, to make a prediction we must plug in the values of the features, so we are assuming we know these values. Going back to the example, to predict profits for the next quarter we must know and plug in the value for weather severity. Weather in which quarter? The current quarter or the next quarter? We might be able to measure the amount of rainfall for this quarter, but if what really matters is next quarter's rainfall, then we must *make a prediction in order to make a prediction*. The moral of this story is that if we care about prediction, we must choose our inputs carefully to avoid this kind of circular dependence.

The second subtle point concerns making predictions for regression and classification problems. In the case of regression, the plug-in method works well since the function $f()$ maps numeric features into a numeric outcome. This is the what we expect of mathematical functions. How, then, can we define functions that output labels of categories, say "dog," "cat," or "train"? The trick used in classification is to use functions that map numeric features into probabilities, numbers between 0 and 1 that quantify our degree of belief that the current example is of a given category. Thus, in classification tasks we generally predict *probability scores*, transformed into categories with the help of a *decision rule*.

Gender Prediction Example

Going back to the gender example that was mentioned when we introduced classification models, suppose that we observe our customers' use of different apps, like Pinterest and Distiller. We want to predict the probability of being a "man" given that we know how many times they use each app per day Prob(Man) = f(times Pinterest, times Distiller). By the properties of probability, the likelihood of not being a man (being a woman or other) is just one minus the probability of being a "man." Acknowledging the strong simplification, for the sake of the discussion let's assume that we only have men or women in our customer base. Generalizing it to the case of multiple genders can be done, but it's easier if we start with a binary classification problem.

Let's say that for one specific customer who visited Pinterest 30 times and Distiller twice, we get 0.51 = Prob(Man) = f(30, 2); in this case the predicted probability of this customer being a "man" is 51% (49% of being a "woman"). The most common decision rule is to assign the category with the largest predicted probability, which in this case turns out to be the male category. This is how all classification models work, including deep learning algorithms for image recognition and natural language processing prediction algorithms. They do not identify categories; they assign probabilities and a decision rule chosen by us, humans, maps them to categories.

Using Supervised Learning as Input for Optimal Decision-Making

I've said throughout that AI is one fundamental input in our decision-making processes. To understand how this works, recall from Chapter 6 that when making decisions under uncertainty, we need to compare the expected value of our metric to be optimized and choose the lever that generates the maximum. The expected value for the case of two consequences is:

$$E(x) = p_1 x_1 + p_2 x_2$$

Classification models provide the probability estimates (p_1, p_2). Alternatively, you may sometimes prefer to use regression directly and estimate the expected utility for each of your levers.

Where Do These Functions Come From?

Now that we know how to make predictions when we have at hand a function that maps inputs into our outcome, the final question is where do these functions come about? Going back to our profits example, how exactly do profits vary with changes in rainfall or labor costs?

The data-driven approach uses supervised algorithms to fit the function that provides the best prediction without imposing any other restrictions. This is by far the most common approach among practitioners. In contrast, the model-driven approach starts from first principles (in some cases without even looking at the data) and imposes restrictions on what type of functions are allowed. This latter approach is mostly used by econometricians and in the industry it's quite uncommon to see. The main advantage of the latter approach is that it provides *interpretable* predictions since theories are built from the ground up. But predictive power is generally lower too.

This trade-off between interpretability and predictive power is pervasive in ML and has several important consequences. First, there may be ethical considerations when we use predictive algorithms to assign offers or make decisions and cannot explain why the decision was made (think of problems like "should I hire that person?" or "is the defendant guilty or not guilty?"). Ethical considerations were discussed in depth in Chapter 8. Second, this lack of transparency can also be a barrier to the widespread adoption of these methods in our businesses. Human beings like to understand *why* things happen before making choices. If data scientists cannot explain *why*, their business counterparts may prefer to keep doing things the traditional way, even if the predictive solution is superior.

Making Good Predictions

By now I hope I convinced you that making predictions is relatively straightforward if we are armed with a function that maps inputs into outcome and we know the values for the features. But how do we make *good* predictions? After all, I can predict that this book is going to be a best-seller, but this can be way off target.

Unfortunately there is no magic recipe for making good predictions. There are, however, some good practices we can follow. The first one is that fitting a function to our data is not the same as making good predictions. As a matter of fact, by memorizing the actual dataset we get a perfect prediction over this data. The key to making good predictions is the ability to *generalize* whenever we receive new data.

Data scientists usually tackle this problem by randomly dividing the dataset into *training* and *test* samples. They first use their algorithms to fit a best function over the training sample and later evaluate the quality of the prediction on the test set. If we focus too much on the training set we might end up *overfitting*, which is a sign of bad generalization in the real world. Notice that in practice we are simulating the "real world" by holding out a subset of the data—the test sample. The quality of this simulation is usually good, unless the world has meaningfully (for the prediction problem) changed between the time we trained the model and time we make a prediction. AI is usually criticized for its lack of ability to generalize in these *nonstationary* settings. We will later encounter an example where overfitting hinders our ability to make good extrapolations to new data.

Another good practice is to ensure that our estimates are robust to small variations of the data. This comes in several flavors. For instance, we can exclude anomalous observations in the training phase as some algorithms are extremely sensitive to outliers.[4] We can also use algorithms that are robust to such datasets, for instance, by averaging or smoothing predictions from different models.

---

4 Averages are extremely sensitive to outliers, as opposed to the median, for example. Take three numbers (1, 2, 3). The simple average and the median are both 2. Change the dataset to (1, 2, 300). The median is still 2, but the average is now 101.

From Linear Regression to Deep Learning

Humans have been using different prediction techniques since the beginning of time, but deep learning is the state of the art and the reason why everybody is currently talking about AI. While this is not the place to delve into the technical details of deep artificial neural networks (ANN), it is nonetheless useful to give an intuitive explanation of how these algorithms work, at least for demystification purposes. To do so we better start with the simplest kind of predictive algorithm: the linear regression.

Linear Regression

In a linear regression we model a quantitative output y as a function of different inputs or features $x_1, x_2, \cdots, x_K$, but we restrict ourselves to linear representations, that is: $y = \alpha_0 + \alpha_1 x_1 + \alpha_2 x_2 + \cdots + \alpha_K x_K$. Notice how the choice of the algorithm restricts the class of functions we can fit to the data. In this case, the learning task is to find weights or coefficients α_k such that we approximate our data well, and most importantly, that we can generalize new data as well.

Imagine that we want to open a new store, but given alternative locations, we want to find the one that maximizes the return on investment (ROI). If we could predict profits for alternative locations, this would be a rather easy task: given our budget, among different locations we would open the store with higher expected profits.

Since profits are revenues minus the cost, the ideal dataset should include these two quantities for all possible locations. As you might imagine, such a dataset may be difficult to come by. Let's say that we were able to get ahold of a dataset that tells us the number of passersby on any given street in our city between 9am and noon. As a starting point, it feels safe to conjecture that more passersby generate higher profits. Our rationale could be as follows: our revenue is derived from sales, which are themselves associated with people that enter our store and end up buying something. The likelihood of making a sale is rather small if no one walks by. On the other hand, if some people happen to pass by, with some luck our shopfront may attract some fraction who might end up buying. This takes care of revenues, but what about costs? It is likely that other store managers think like us, so we may have to compete for the rent, thereby increasing our costs. Notice that this story only talks about what we believe are plausible *directions* of the effects and not the actual *magnitudes*; we will let the algorithms fit the data and check whether the results match our intuition.

A simple linear model could then pose that profits (y) are increasing with the number of passersby (x_1), or $y = \alpha_0 + \alpha_1 x_1$. Note that we, the human experts, have decided three things: (1) the outcome we want to predict is profits for each store; (2) profits depend *only* on the number of passersby at each location, a notable simplification; and (3) we restrict ourselves to a linear world, another notable simplification.

In linear regression we then ask: what are the parameters α_0, α_1 that make our *prediction most accurate?*

Figure A-2 shows a scatterplot of profits as function of passersby for 10 hypothetical stores currently operating for our company. We can immediately see that there is an apparent trend in the data as higher profits are associated with more passersby.

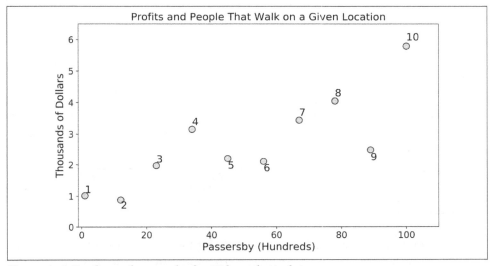

Figure A-2. Profits and passersby for 10 hypothetical stores

The least-squares method is the most common solution of what we mean by "the prediction being most accurate." Under this loss function the algorithm finds parameters that minimize the average error, as seen in Figure A-3.[5] The prediction under this linear model is given by the dashed line, and errors—depicted with vertical lines—can be either positive or negative. For store 10, for example, our prediction is almost $2K dollars below the actual number, and the error is positive. For stores 2, 5, 6, and 9, on the other hand, our linear model overestimates actual profits given their number of passersby. At the bottom of the plot the actual predicted equation is presented. For this dataset, the algorithm found that $\alpha_0 = 0.9$ and $\alpha_1 = 0.04$. The details of how this happened are beyond the scope of this book, but let me just point out that the algorithm tries different possible values in a way that reduces the amount of computation.

---

5 Formally, it minimizes the mean squared error. That way we treat positive and negative errors symmetrically.

A naive alternative is to try many different random possible combinations, but since this set is huge, it would take years for us to find the best possible values.[6]

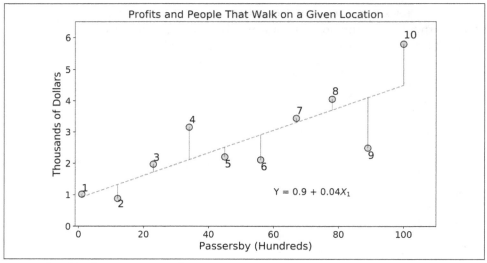

Figure A-3. Prediction errors in linear regression

This equation is not just an intellectual curiosity for the data scientists that took pains to compute it; it actually provides some interpretation of the plausible mechanism that converts passersby into profits: each additional 100 people that walk through the location "generate" $40 in profits. We discussed the risks of interpreting these results causally in Chapter 2, but for now it is important to highlight that this mathematical object may have substantial business content, the type of content necessary to create value using AI, as it allows us to interpret our results in a very simple way. As discussed previously, linear regression is on one extreme of the predictive power versus interpretability spectrum, making it a common choice of algorithm if understanding and explainability are sufficiently appreciated.

Going back to our results, notice that the prediction is really good for some stores and rather lousy for others: our linear model gave almost perfect predictions for stores 1, 3, 5, 7, and 8, but did poorly for stores 9 and 10. One property of the least

---

6 The method of ordinary least squares (OLS) actually has a closed-form solution, meaning that in practice we do not have to iterate until converging to a good-enough estimate. There are setups, however, where iterative improvements are made by using *gradient descent* that sequentially adjusts the weights in such a way that we approach the set of optimal parameters (Chapter 7). This is usually the case when we use OLS with big data where we end up distributing the data across several nodes or servers.

squares solution is that the average error is zero,[7] so positive and negative errors end up balancing out.

Why Linear Regression

One final detail that causes confusion in linear regression is what "linear" really means. Figure A-4 shows the result of fitting a linear regression with quadratic effects. Suppose that we suspect that revenues grow faster than costs as a function of the number of potential customers, so it is likely that profits may accelerate nonlinearly. A linear regression can deal with such nonlinear effects without a problem: "linearity" pertains to how the unknown parameters enter into the equation, not whether our inputs are nonlinear. In this case we included a second term, the square of the number of passersby, to capture this potential nonlinearity, and the algorithm was perfectly capable of doing so.

Other nonlinearities that are common are interaction effects. Imagine that we now have a second variable representing the average household income in the store's neighborhood. Our product, however, is targeted to lower-to-medium-income customers. We might therefore suspect that in medium-to-upper-income neighborhoods each additional passerby creates less profits compared to our target lower-medium-income neighborhoods. This is an example where to have a good prediction model, we need to let some of our features interact with each other.

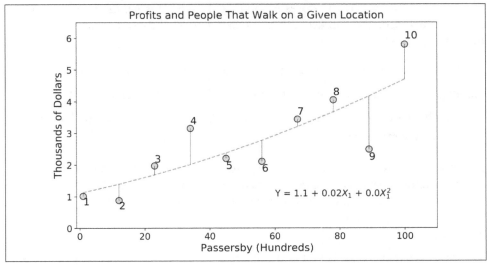

$$Y = 1.1 + 0.02X_1 + 0.0X_1^2$$

Figure A-4. Fitting a quadratic equation

---

[7] This is true whenever we include a constant, intercept, or bias, α_0.

Controlling for other variables

Linear regression is great to capture some intuition about relationships found in the data. In my opinion, one of the most important results in linear regression is known as the *Frisch–Waugh–Lovell theorem* (FWL) (*https://oreil.ly/WHtde*) and states that by including control variables (confounders in the language of Chapter 2) in the regression, we estimate the net effect of a variable of interest.

To see this in action, suppose that you suspect that the results in the previous regression may be contaminated by the fact that there are more potential customers in highly commercial neighborhoods, so your lower profits may be related to price effects in highly competitive neighborhoods and not to the sheer volume of customers. This third variable could act as a confounder of the pure volume effect you were trying to estimate.

You can control for this effect by including the extra variable—number of commercial establishments in the neighborhood—in your linear regression:

$$y_i = \alpha_0 + \alpha_1 x_{1i} + \alpha_2 x_{2i} + \epsilon_i$$

As before, x_{1i} denotes the number of passersby per period of time and x_{2i} represents the number of different stores in the neighborhood, both for store i.

The FWL theorem says that if you run this regression, the estimated coefficient for the number of passersby (or stores in the neighborhood) is *net* of any other control variables included. You can actually check that you get the *exact same result* by estimating the following three alternative regressions:

1. Regress y_i on x_{2i} (and a constant) and save the residuals (call them η_i).
2. Regress x_{1i} on x_{2i} (and a constant) and save the residuals (call them v_i).
3. Finally, regress η_i on v_i (and a constant). The corresponding coefficient for v_i is exactly the same as the estimated one from the longer regression $\hat{\alpha}_2$.

Note that in steps 1 and 2 you're *cleaning* the variables of interest of the impact that the confounder may have. Once you have these new variables (the corresponding residuals), you can run a simple bivariate regression to get the desired estimate. The book's Github repository provides a demo of the theorem in action as well as some additional details.

The very nice conclusion is that you can get net effects by including the controls or confounders in the same regression.

Overfitting

Figure A-5 shows the result from fitting a linear regression that includes a polynomial of order 6. The top panel shows that this function fits the training data outstandingly, don't you think? However, remember that we want to generalize to new data. The bottom panel shows the prediction on our test data consisting of 10 new stores that were *not* used in the training stage. It appears that the fit is not as good as on the test set, though. To check for overfitting we would actually need to plot the prediction error for both the training and test data as a function of the degree of the polynomial.

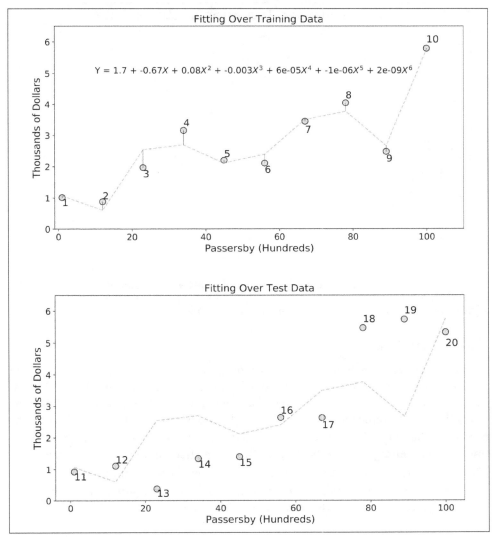

Figure A-5. A case of overfitting

Figure A-6 depicts the average prediction error for both the training and test data, as a function of the flexibility we allow in the linear regression in terms of the degree of the polynomial. As you can see, in the training data we are always doing a little better, with errors consistently falling with higher degrees. This is always the case with training data and is the reason why data scientists evaluate the quality of their prediction on another set. For test data, there is evidence of overfitting with polynomials of degree 3 or higher: the lowest prediction error is obtained with polynomial of degree 2, and afterward it starts increasing, showing the lower ability to generalize.

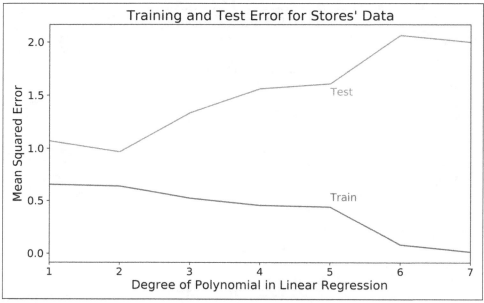

Figure A-6. Prediction error for training and test samples

Neural Networks

To introduce neural networks, let me start by drawing a somewhat peculiar diagram for linear regression just for the purpose of building a bridge between the two techniques. Figure A-7 shows circular nodes that are connected though edges or lines; the strength of each connection is given by weights α_k. This collection of nodes and the corresponding edges is called a network, and it provides a way to visualize the relationship between inputs and the output: X_1 has an effect on the output of magnitude α_1, and similarly for X_2. In the discussion that follows, keep in mind that the learning algorithm in the case of linear regression finds the set of weights that minimize the prediction error, as *this will also be the case for deep neural nets.*

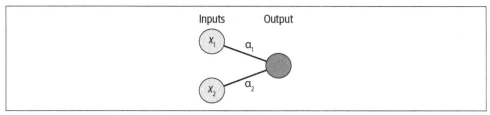

Figure A-7. Linear regression depicted as a network

In regression, since $y = \alpha_0 + \alpha_1 X_1 + \alpha_2 X_2$ the quantity on the righthand side maps one-to-one to outcome y. For instance, if $\alpha_1 = 1$, $\alpha_2 = 0$ and there is no intercept or bias, if the corresponding inputs have values 2 and 1, thanks to the equality, we know that $y = 2$. This is one of the properties that linearity buys for us. But we can imagine having something more general, say $y = g(\alpha_0 + \alpha_1 X_1 + \alpha_2 X_2)$ for some function $g()$. This would most likely take us out of the realm of linear regression, but depending on the problem it may fit the data better, and hopefully also generalize better to new data.[8]

Activation functions: adding some extra nonlinearity

Activation or *transfer* functions are one such class of functions, and are commonly used in deep learning (DL). The implicit activation function in linear regression is, of course, linear (see Figure A-8). But there are several more interesting alternatives that have become essential for practitioners. In the rectified linear unit (ReLU) case, output is zero when $\alpha_0 + \alpha_1 X_1 + \alpha_2 X_2$ is negative or zero, and for positive values we are again in the world of linear regression. Notice how the function gets activated only when the joint effort is positive, hence its name. A smoother version is the sigmoid activation. Activation functions are another way to include nonlinear effects in a predictive model. They were first introduced to capture our understanding of how neurons in our brain fire, but today we use them because they improve the predictive power of our models.

---

8 In some cases we can still use linear regression, say $g(z) = exp(z)$. Using the logarithm transformation takes us back to the linear realm with a new, transformed outcome ln (y).

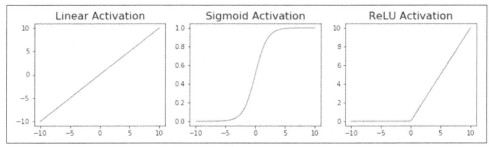

Figure A-8. Different transfer or activation functions

With these tools we are ready to introduce neural networks. A neural network is a set of nodes and edges or connections between the nodes like the one depicted in Figure A-9. The left panel shows a network with only two inputs, each of which affects two hidden or intermediate nodes, the strength mediated by the corresponding parameters. We may decide to use nonlinear activation functions in each of the intermediate nodes. Finally, once we have the strength of each hidden unit we are ready to aggregate the impact onto the output, mediated again by corresponding weights and possibly an activation function. It may not be immediately obvious, but these arrangements of nodes and edges were originally designed to emulate how our brains work, nodes being the neurons that are connected through synapses or edges in the network. These days, few practitioners take this analogy literally. The right panel shows a somewhat deeper network with two hidden layers, each one with different widths or number of nodes. It is fully connected since every node in a layer is connected to nodes in the subsequent layer, but this need not be the case. As a matter of fact, one way to control for overfitting in DL is by systematically deleting some of these edges between nodes, a technique known as *dropout*.

Deep ANNs keep growing larger, with very deep nets having hidden layers in the hundreds and neurons in the hundreds of thousands and even millions. Moreover, richer architectures or ways to assemble these neurons have proved invaluable in solving specific problems in computer vision and natural language processing, and this is today an active line of research.

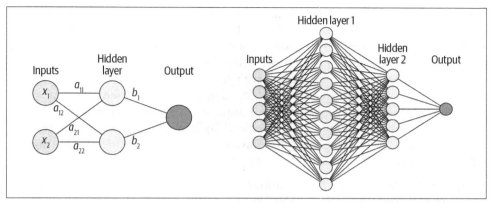

Figure A-9. Two examples of deep neural nets: the left plot shows a network with one hidden layer and two inputs, while the plot on the right depicts a network with five inputs and two hidden layers

The success of deep learning

So what is the buzz around DL? ANNs and their distinct architectural variations, such as convolutional neural networks (CNNs) and the like, have proven to be highly successful at solving problems that were previously thought to be solvable at the human level only by us.

In the field of image recognition, deep CNNs have achieved results comparable to those of humans, and have even surpassed our species in image classification. In 2012, AlexNet, one such neural net, showed DL's power relative to other approaches by winning the ImageNet competition and reducing the classification error from 26% to 15.3%.[9] This error has kept decreasing every year with deeper networks, now even beating our capacity to recognize and classify objects in up to 1,000 categories.[10] Each layer in a deep CNN detects different levels of abstractions and details in an image. For instance, the first layer may detect the shapes or edges of the objects, the second layer patterns such as lines or stripes, and so on.

Language is an example of sequence data where RNNs have had a profound impact. Sequential data has the property that the order of occurrence matters. For instance, take these two variations: "I had my room cleaned" and "I had cleaned my room." The two sentences have the exact same words in different order but slightly different meanings. Another example of how order matters is the prediction of one word given

---

9 These are top-5 error rates, meaning that we use the top 5 predictions for each image, and check if the real label is included in this list. Top-1 error rate is the traditional prediction error: algorithms are allowed only one guess per image. Human error rates on this dataset have been benchmarked to around 3.6%, meaning that out of 100 images to be classified, our top 5 predictions match almost 96 of them.

10 The ImageNet database has more than 20,000 categories, but only 1,000 are used in the competition.

the neighboring words, a process analogous to the way we extract context in a sentence. One difficulty with sequence data is that the algorithm must have some type of memory that remembers what happened previously and what the current state is. In contrast to the multilayer neural nets, when successive inputs are used, an RNN accesses this memory and combines it with the current input to update the prediction. This class of networks has shown incredible power in problems such as speech recognition, image captioning, machine translation, and question answering, and is behind every virtual assistant that is commercially available. Thanks to these technologies, companies are transforming the ways we interact and communicate with our customers. This is only the tip of the iceberg for natural language applications.

Another area where DL has had a profound effect is video games and games of strategy playing. Starting in 2015, Google's DeepMind AlphaGo has beaten human professionals and top players in the game of Go. This is done by combining the predictive power of DL algorithms with *reinforcement learning*. In the latter, a well-defined system of rewards operates when certain actions are taken. The task of the algorithm is to learn what actions earn the highest rewards possible into the future by interacting with the environment, and good or bad decisions generate a flow of rewards that reinforce the learning process. At this point, we have only been able to use these kinds of algorithms in highly constrained settings, such as games where the reward function is relatively simple and where we can easily generate massive datasets of play for the algorithms to learn. But researchers are already using these technologies to improve autonomous or self-driving vehicles, robots, and factories, so the future seems promising for deep reinforcement learning.

While DL (and ML more broadly) has and will keep having a profound impact, it is not a general-purpose learning technology (like the one humans have). Some of the areas where ML doesn't fare well compared to humans are the identification and learning of causal relationships, learning with little experience (semisupervised or self-supervised learning), common-sense calculations, and context extraction, just to name a few. Machine learning algorithms are powerful pattern-recognition techniques and should be treated as such.

A Primer on A/B Testing

The previous sections introduced some basic concepts in ML, with emphasis on supervised learning techniques that are most commonly used by practitioners. In this last section we will take a look at A/B testing, the technique that was introduced in Chapter 2.

Recall that our objective here is to simulate counterfactuals that eliminate the pervasive effect that selection bias has. The idea is that while we may not be able to get exact copies of our customers, we may still be able to simulate such a copying device using *randomization*, that is, by randomly assigning customers to two groups: those who receive the treatment and those who don't (the control group). Note that the choice of two groups is done for ease of exposition, as the methodology applies for more than two treatments.

We know that customers in each group are different, but by correctly using a random assignment we dispose of any selection bias: our customers were selected by chance, and chance is thought to be unbiased. Good randomization buys for us that the treatment and control groups are *ex-ante indistinguishable* on average. To give a more precise idea of how this works, suppose that we know the age and gender of our customers. *Before running the experiment* we would check if the gender distribution and average ages are the same in the treatment and control groups. If the answer is positive, then our randomization was done correctly.

Risks When Running Randomized Trials

We have noted that randomization is unbiased in the sense that the result of a random draw is obtained by chance. In practice we simulate pseudorandom numbers that have the look and feel of a random outcome, but are in fact computed with a deterministic algorithm. For instance, in Excel, you can use the =RAND() function to simulate a pseudorandom draw from a uniform distribution.

It is important to remember, however, that using randomization does not necessarily eliminate selection bias. For example, even though the probability of this happening may be *extremely* low, by pure chance we may end up with a group of male customers receiving the treatment and females in the control group, so our random assignment ended up selecting by gender, potentially biasing our results. It's a good practice to check if random assignment passes the ex-ante test by checking differences in means on observable variables.

Last but not least, there may be ethical concerns since in practice we are potentially affecting the outcomes of one group of customers. One should always checklist any ethical considerations you might have before running an experiment. You can refer to Chapter 8 for a more comprehensive discussion on ethical issues in A/B testing.

A/B testing in practice

In the industry, the process of randomizing to assign different treatments is called A/B testing. The name comes from the idea that we want to test an alternative *B* to our default action *A*, the one we commonly use. As opposed to many of the techniques in the machine learning toolbox, A/B testing can be performed by anyone, even without a strong technical background. We may need, however, to guarantee that our test satisfies a couple of technical statistical properties, but these are relatively easy to understand and put into practice. The process usually goes as follows:

1. Select an actionable hypothesis you want to test: for example, female call center representatives have a higher conversion rate than male representatives. This is a crisp hypothesis that is falsifiable.

2. Choose a relevant and measurable KPI to quantify the results from the test; in this example, we choose conversion rates as our outcome. If average conversion for female reps *isn't* "significantly larger" than that for men, we can't conclude that the treatment worked, so we keep running the business as usual. It is standard practice to use the concept of statistical significance to have a precise definition of what *larger* means.

3. Select the number of customers that will be participating in the test: this is the first technical property that must be carefully selected and will be discussed next.

4. Randomly assign the customers to both groups and check that randomization produced groups that satisfy the ex-ante indistinguishable property.

5. After the test is performed, measure the difference in average outcomes. We should take care of the rather technical detail of whether a difference is generated by pure chance or not (statistical significance).

If randomization was done correctly, we have eliminated the selection bias, and the difference in average outcomes provides an estimate of the causal effect.

Understanding power and size calculations

Step 3, selecting the number of customers, is what practitioners call *power and size calculations*, and unfortunately there are key trade-offs we must face. Recall that one common property of statistical estimation is that the larger the sample size, the lower the uncertainty we have about our estimate. We can always estimate the average outcome for groups of 5, 10, or 1,000 customers assigned to the B group, but our estimate will be more precise for the latter than for the former. From a strictly statistical point of view, we prefer having large experiments or tests.

From a business perspective, however, testing with large groups may not be desirable. First, our assignment must be respected until the test comes to an end, so there is the opportunity cost of trying other potentially more profitable treatments, or even our control or base scenario. Because of this, it is not uncommon that the business stakeholders want to finish the test as quickly as possible. In our call center example, it could very much have been the case that conversion rates were *lower* with the group of female reps, so during a full day we operated suboptimally, which may take an important toll on the business (and our colleagues' bonuses). We simply can't know at the outset (but a well-designed experiment should include some type of analysis of this happening).

Because of this trade-off we usually select the *minimum* number of customers that satisfies two statistical properties: an experiment should have the right statistical size and power so that we can conclude with enough confidence if it was a success or not. This takes us to the topic of false positives and false negatives.

False positives and false negatives

In our call center example, suppose that contrary to our initial assumption, male and female representatives have the exact same conversion efficiency. In an ideal scenario we would find no difference between the two cases, but in practice this is *always* non-zero, even if small. How do we know if the difference in average outcomes is due to random noise or if it is showing a real, but possibly small difference? Here's where statistics enter the story.

There is a *false positive* when we mistakenly conclude that there is a difference in average outcomes across groups and we therefore conclude that the treatment had an effect. We choose the *size of the test* to minimize the probability of this happening.

On the other hand, it could be that the treatment actually worked but we may not be able to detect the effect with enough confidence. This usually happens when the number of participants in the experiment is relatively small. The result is that we end up with an *underpowered* test. In our call center example, we may falsely conclude that representatives' productivity is the same across genders when indeed one has higher conversion rates.

Statistical Size and Power

Somewhat loosely speaking, the *size* of a statistical test is the probability of encountering a false positive. The *power* of the test is the probability of correctly finding a difference between treatment and control.

The first panel in Figure A-10 shows the case of an underpowered test. The alternative B treatment creates 30 additional sales, but because of the small sample sizes, this difference is estimated with insufficient precision (as seen by the wide and overlapping confidence intervals represented by the vertical lines).

The second panel shows the case where the real difference is close to 50 extra sales, and we were able to precisely estimate the averages and their differences (since confidence intervals are so small that they don't even look like intervals).

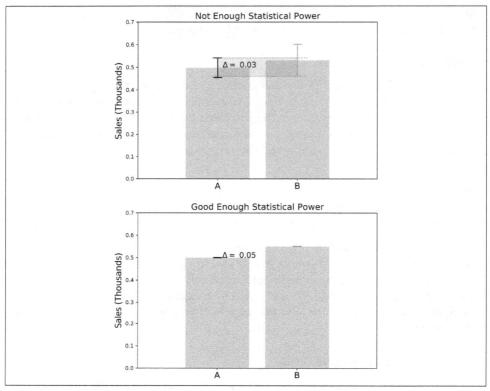

Figure A-10. The first panel shows the result of an underpowered test: there is a differ-ence in the average outcomes for the treated and untreated, but the small sample sizes for each group cannot estimate this effect with enough precision; the second panel shows the ideal result where there is a difference and we can correctly conclude this is the case

Let's briefly talk about the costs of false positives and false negatives in the context of A/B testing. For this, recall what we wanted to achieve with the experiment to begin with: we are currently pulling a lever and want to know if an alternative is superior for a given metric that impacts our business. As such, there are two possible out-comes: we either continue pulling our A lever, or we substitute it with the B alterna-tive. In the case of a false positive, the outcome is making a subpar substitution. Similarly, with a false negative we mistakenly continue pulling the A lever, which also impacts our results. In this sense both are kind of symmetric (in both cases we have an uncertain long-term impact), but it is not uncommon to treat them asymmetri-cally, by setting the probability of a false positive at 5% or 10% (size), and the proba-bility of a false negative at 20% (one minus the power).

There is however the opportunity cost of designing and running the experiment, so we'd better run it assuming the best-case scenario that the alternative has an effect.

That's why most practitioners tend to fix the size of a test and find the minimum sample size that allows us to detect some minimum effect.

Selecting the Sample Size

In tests where we only compare two alternatives, it is common to encounter the following relationship between the variables of interest:

$$MDE = \left(t_\alpha + t_{1-\beta}\right)\sqrt{\frac{\text{Var(Outcome)}}{NP(1-P)}}$$

Here t_k is a critical value to reject a hypothesis with probability k according to a t distribution, α and $1 - \beta$ are the size and power of the test (that you can replace to calculate corresponding critical values), MDE the minimum detectable effect of the experiment, N the number of customers in the test, P is the fraction assigned to the treatment group, and Var(Outcome) is the variance of the outcome metric you're using to decide if the test is successful or not.

As you can see from this formula, for a given MDE the larger the variance of your outcome, the larger the sample you will need. This is standard in A/B testing: noisy metrics will require larger experiments. Also, remember that our objective is to have a small enough MDE that allows us to detect incremental changes caused by the treatment, putting even more pressure on the size of the experiment. You can find a full derivation of this equation on the Github repository for the book.

Example A-1 shows how to calculate the sample size for your experiment with Python:

Example A-1. Calculating the sample size for an A/B test

```
# Example: calculating the sample size for an A/B test
from scipy import stats
def calculate_sample_size(var_outcome, size, power, MDE):
    '''
    Function to calculate the sample size for an A/B test
    MDE = (t_alpha + t_oneminusbeta)*np.sqrt(var_outcome/(N*P*(1-P)))
    df: degrees of freedom when estimating the variance of the outcome
    (if sample size is large df is also large so I artificially set it at
    1000)
    '''
    df = 1000
    t_alpha = stats.t.ppf(1-size, df)
    t_oneminusbeta = stats.t.ppf(power, df)
    # same number of customers in treatment and control group
    P = 0.5
```

```
    # solve for the minimum sample size
    N = ((t_alpha + t_oneminusbeta)**2 * var_outcome)/(MDE**2 * P * (1-P))
    return N

# parameters for example below
var_y = 4500
size = 0.05
power = 0.8
MDE = 10
sample_size_for_experiment = calculate_sample_size(var_y, size, power, MDE)
print('We need at least {0} customers in experiment'.format(
    np.around(sample_size_for_experiment),decimals=0))
```

In practice, we start by setting the power and size of the test and then choose an MDE. One way to think about it is that it is the minimum change on our outcome metric that makes the experiment worthwhile from a business standpoint. We can finally reverse engineer the sample size we need from the formula.

To see this in practice, suppose that we want to run an A/B test to see if we can increase our average customer spend or ticket by way of a price discount. In this price elasticity experiment, the treatment group will get the new lower price, and the control will keep paying the regular price. Because of those very high-spend customers, the variance in monthly spend is $4,500 (standard deviation is about $67). As a benchmark we choose standard values for size and power (5% and 80%). Finally, our business stakeholders convince us that from their perspective it only makes sense to try the new alternative if we find a minimum effect (MDE) of 10 dollars (or 15% of one standard deviation). We run our size calculator and find that we need at least 1,115 participants in the experiment. Since our contact rate is around 2%, we should send emails to around 1115/0.02 = 55.2K customers.

Further Reading

There are many books on general machine learning. Readers looking for highly technical material can always go to Robert Tibshirani's and Trevor Hastie's masterpiece *Elements of Statistical Learning* (Springer). Together with Daniela Witten and Gareth James, they also authored an introductory textbook on building and interpreting machine learning models that is less technical and rich on intuition: *Introduction to Statistical Learning With Applications in R* (Springer). Kevin Murphy's *Machine Learning: A Probabilistic Perspective* (MIT Press) is great at presenting many methods at a technical level as well as providing much-needed Bayesian foundations.

Other books that are more hands-on, and will teach you not only to understand but also to implement machine learning with popular open source libraries are Matthew Kirk's *Thoughtful Machine Learning* (O'Reilly) and Joel Grus's *Data Science from Scratch* (O'Reilly). They both provide superb introductions to machine learning models, as well as in-depth discussions into some of the methods that are commonly used.

I also highly recommend Foster Provost and Tom Fawcett's *Data Science for Business* (O'Reilly), a book directed towards business people and practitioners alike: they achieve the hard-to-get balance between the technical and explanatory. These last three are all part of the highly-recommended O'Reilly series on machine learning and data science.

Probably the most thorough and general treatment of artificial neural networks can be found in *Deep Learning* by Ian Goodfellow, Yoshua Bengio, and Aaron Courville (MIT Press). I also recommend *Deep Learning* by Adam Gibson and Josh Patterson (O'Reilly), as well as Francois Chollet's *Deep Learning With Python* (O'Reilly). There are also advanced books for specific topics: Alex Graves's *Supervised Sequence Labelling with Recurrent Neural Networks (Studies in Computational Intelligence)* (Springer) provides a thorough treatment of RNNs, and Yoav Goldberg's *Neural Network Methods in Natural Language Processing (Synthesis Lectures on Human Language Technologies)* (Morgan & Claypool) is an introduction to learning algorithms in natural language applications.

Two surprisingly refreshing general-public treatments of current AI technologies are Terrence Sejnowski's *The Deep Learning Revolution* (MIT Press) and Jerry Kaplan's *Artificial Intelligence. What Everyone Needs to Know* (Oxford University Press). Written by an insider in the revolution, though highly self-referential, the former provides a rich chronological account of the development of the techniques in the last decades of the last century, as well as how what neuroscientists and cognitive scientists know about the workings of the brain motivated many of those developments. The latter's Q&A format may seem too rigid at times, but in the end it provides an easy-to-read account of current methods, as well as less common topics such as the philosophical underpinnings behind AI.

There are many treatments of A/B testing, starting with Dan Siroker and Pete Koomen's *A/B Testing: The Most Powerful Way to Turn Clicks into Customers* (Wiley) where you can get a good sense of A/B testing applications. Peter Bruce and Andrew Bruce's *Practical Statistics for Data Scientists* (O'Reilly) provides an accessible introduction to statistical foundations, including power and size calculations. Carl Andersen's *Creating a Data-Driven Organization* (O'Reilly), briefly discusses some best practices in A/B testing, emphasizing its role in data- and analytics-driven organizations. Ron Kohavi (previously at Microsoft and now at Airbnb) has been forcefully advancing the use of experimentation in the industry. His recent book, coauthored with Diane Tang and Ya Xu, *Trustworthy Online Controlled Experiments: A Practical Guide to A/B Testing* (Cambridge University Press) is a great reference. Some of this material can be found at *https://exp-platform.com* and *https://oreil.ly/1CPRR*.

You can find material on power calculations in Howard S. Bloom's *The Core Analytics of Randomized Experiments for Social Research*, available online (*https://oreil.ly/zs9gx*).

Index

AI data bias sensitivity, 54
    gender prediction example, 195
generalizations
    good prediction practices, 197
    overfitting, 197, 203
    semisupervised learning, 193
genetics in human levers, 69
    behavioral genetics, 70
Giffen goods, 67
Google
    big data distributed computation, 5
    business questions first, 42
grade of service (GoS), 172
gradient descent optimization, 148-150, 200

H

hard metrics, 41
    personnel example, 54, 138
herding behavior, 72
heterogeneity
    causal effect estimation, 25
    human levers, 65
    simplifying to homogeneity, 98
    uncertainty from, 33, 106, 171
heuristics in decision-making, 76
hiring personnel use case
    business question framing, 53
    employee lifetime value, 54, 138
    levers for, 78
    team lifetime value, 139
    uncertainty, 138
history
    big data, 5
    online advertising importance, 8
human levers, 64-72
    behavioral levers, 79
    books on, 80
    conformity and peer effects, 72
    expectation changers, 76
    framing effects, 73
    genetics, 69
    individual and social learning, 70
    loss aversion, 74
    preferences in, 69-75
    pricing lever, 66
    social influences, 71
    time constraints, 69
    two-sided platform adoption, 71
    why do humans behave as we do, 65-72

I

ignorance causing uncertainty, 36, 106
image recognition via deep learning, 207
ImageNet competition, 207
influencers, 71, 72, 72
information overload at customers, 156
inventory optimization use case
    books on, 104
    business question framing, 56
    levers for, 79
    simplification of, 102
    uncertainty, 140

K

key performance indicators (KPIs)
    business objective evaluation, 41
    CAPEX optimization, 51, 78, 136
    continuous and regression, 131
    cross-selling, 49, 77, 135, 171
    customer churn, 47, 77, 133
    customer lifetime value, 17, 48
    customer satisfaction, 134
    dating app, 44
    delinquency rates, 55, 79, 139
    employee lifetime value, 54, 138
    hard versus soft metrics, 41
    hiring personnel, 53, 78, 138
    inventory optimization, 56, 79
    loyalty program, 18
    selecting, 18
    sequence of why questions for, 19
    store location, 52, 78, 137, 169
    store staffing numbers, 57, 79, 163, 171
    team lifetime value, 139
Kozyrkov, Cassie, 42
KPI (see key performance indicators)

L

language and RNNs, 207
    books on, 215
Law of Demand, 66
    Giffen goods, 67
    optimization of price, 151
    pricing lever as constraint, 69
least-squares method
    average error zero, 200
    ordinary least squares, 200
    prediction most accurate, 199

statistics
  books on, 141, 215
  confidence intervals, 132
  false positive, 211
  false positive/negative costs, 212
  Occam's razor, 23
  power of statistical test, 211
  probabilities as objective truth, 131
    (see also probability theory)
  sample size, 210, 213
  size of statistical test, 211
  statistical significance for evaluation, 210
stock optimization use case
  books on, 104
  business question framing, 56
  levers for, 79
  simplification of, 102
  uncertainty, 140
store location use case
  business question framing, 52
  levers for, 78
  optimization without uncertainty, 169
  simplification of, 99
  uncertainty, 137
store staff numbers use case
  business question framing, 57
  Erlang C model, 171
  levers for, 79
  optimization objective function, 163, 168
  optimization with uncertainty, 171
  optimization without uncertainty, 163-169
  simplification of, 103
strategic behavior books, 81
structured data, 6
summary of book
  causality, 181
  optimization, 184
  prescriptive questions, 179
  simplification, 182
  uncertainty, 182
supervised learning
  about, 191
  classification, 127, 128, 194, 195
  learning how to make decisions, 186
  optimization, 146
  optimization objective function, 184
  regression, 127, 131, 193
  semisupervised learning, 193

T

Taylor series approximation, 85
team lifetime value, 139
technology as physical lever, 63
teens and conformity, 72
tennis balls Fermi problem, 86
The Analytical Mind (Rutherford), 9
theory of expected utility, 121
thinking outside the box, 181
transfer functions (see activation functions)
Traveling Salesman optimization problem, 146
two-sided platforms, 44
  books on, 59
  dating app decomposition, 44
  malls as, 138
  social effects in adoption, 71

U

uncertainty
  bandit problems, 130
  books on, 141
  CAPEX optimization with, 136
  CAPEX optimization without, 157-163
  credit card offer user case, 136
  cross-selling, 135
  cross-selling optimization with, 171
  cross-selling optimization without, 155
  customer churn, 132
  customer churn optimization with, 170
  customer churn optimization without, 153
  in decision-making, 31-36, 76, 105, 116-121,
    126, 169
  decision-making paradoxes, 121
  decision-making without, 112, 118, 120, 182
  decisions simple but uncertain, 114
  delinquency rates, 139
  Game of Life, 35
  heterogeneity causing, 33, 106, 171
  hiring personnel, 138
  ignorance causing, 36, 106
  inventory optimization, 140
  lottery paradox, 121
  modeling, 134
  optimization with, 169-177
  optimization with, tricks for, 174
  optimization without, 153-170
  quantifying, 106-112
  risk aversion, 114
  sample size and, 210

About the Author

Daniel Vaughan currently leads the data science function for Airbnb in Latin America. Previously he was Chief Data Officer and Head of Data Science at Telefonica Mexico. As CDO, he oversaw the design and monitoring of the company's big data strategy, from technical decisions such as the data lake implementation and optimization, to strategic aspects such as value-based ingestion and governance of all the relevant data sources. As Head of Data Science, he led a team of data scientists and engineers responsible for developing all predictive and prescriptive solutions in Mexico. An important part of this role was to work across the board with business stakeholders, helping them translate business questions into problems that could be solved through AI and data science. Before, he held positions as Senior Data Scientist at Banorte, and Economic Researcher at Banco de Mexico. He is passionate about finding new ways to translate business problems into prescriptive solutions, as well as strategies to speed up the transition to a data-driven culture. He holds a Ph.D. in economics from New York University (NYU) and has taught technical and non-technical courses at NYU (USA and UAE), Universidad de los Andes and Universidad ICESI (Colombia), CIDE and TEC de Monterrey (Mexico). He regularly lectures on the material of this book to MBA students at EGADE Business School in Mexico City.

Colophon

The animal on the cover of *Analytical Skills for AI and Data Science* is the hooded crow (*Corvus cornix*). This crow, also known as the Scotch crow and the Danish crow (among other local monikers), can be found all across northern and eastern Europe, including parts of the British Isles, as well as in some parts of northwest Asia.

The hooded crow's plumage is primarily grey, while its head, throat, wing, and tail feathers are black. While male hooded crows tend to be larger than females, they are not otherwise sexually dimorphic. Juvenile hooded crows, however, display duller plumage and different coloration in the eyes and mouth. A constant scavenger, the hooded crow is omnivorous and has a similar diet to that of its closely-related cousin, the carrion crow, consisting of small mammals, shellfish, eggs, nuts, scraps, and so forth. The hooded crow will also hide its food in places such as gutters, flower pots, or under the cover of bushes to eat later. Other crows may often keep an eye out for these hiding places so that they can steal the food later.

The hooded crow is sometimes grouped along with the carrion crow as a single species, and hybrids of the two birds are not uncommon. It appears in Celtic and Faroese folklore, and is one of the 37 Norwegian birds depicted in the Bird Room of the Royal Palace in Oslo.

While its current conservation status is of "Least Concern," many of the animals on O'Reilly covers are endangered; all of them are important to the world.

The cover illustration is by Karen Montgomery, based on a black and white engraving from *British Birds*. The cover fonts are Gilroy Semibold and Guardian Sans. The text font is Adobe Minion Pro; the heading font is Adobe Myriad Condensed; and the code font is Dalton Maag's Ubuntu Mono.

O'REILLY®

There's much more where this came from.

Experience books, videos, live online training courses, and more from O'Reilly and our 200+ partners—all in one place.

Learn more at oreilly.com/online-learning

©2019 O'Reilly Media, Inc. O'Reilly is a registered trademark of O'Reilly Media, Inc. | 175

CPSIA information can be obtained
at www.ICGtesting.com
Printed in the USA
JSHW031145230520
5863JS00001B/1